STUDENT UNIT GUIDE

AS Geography
UNIT 2

Specification A

Module 2681: The Human Environment

Peter Stiff

My thanks to Alison, Phillip and Timothy for their help and encouragement

Philip Allan Updates, Market Place, Deddington, Oxfordshire, OX15 0SE

Orders
Bookpoint Ltd, 130 Milton Park, Abingdon, Oxfordshire, OX14 4SB
tel: 01235 827720   fax: 01235 400454
e-mail: uk.orders@bookpoint.co.uk
Lines are open 9.00 a.m.–5.00 p.m., Monday to Saturday, with a 24-hour message answering service. You can also order through the Philip Allan Updates website: www.philipallan.co.uk

© Philip Allan Updates 2002

ISBN-13: 978-0-86003-743-9
ISBN-10: 0-86003-743-6

All rights reserved; no part of this publication may be reproduced, stored in a retrieval system, or transmitted, in any form or by any means, electronic, mechanical, photocopying, recording or otherwise without either the prior written permission of Philip Allan Updates or a licence permitting restricted copying in the United Kingdom issued by the Copyright Licensing Agency Ltd, 90 Tottenham Court Road, London W1T 4LP.

In all cases we have attempted to trace and credit copyright owners of material used.

Exam questions are reproduced by permission of OCR.

This Guide has been written specifically to support students preparing for the OCR Specification A AS Geography Unit 2 examination. The content has been neither approved nor endorsed by OCR and remains the sole responsibility of the author.

Typeset by TLC Design Management
Printed by MPG Books, Bodmin

Philip Allan Updates' policy is to use papers that are natural, renewable and recyclable products and made from wood grown in sustainable forests. The logging and manufacturing processes are expected to conform to the environmental regulations of the country of origin.

Contents

Introduction
About this guide 4

The aim of Module 2681 4

Examination skills 5

■ ■ ■

Content Guidance
About this section 10

Population

Population distribution 11

Population change through time 16

Population change through space 19

Rural settlement

Rural settlements in MEDCs 22

Population change in MEDCs since 1960 29

Urban settlement

Contemporary urbanisation in LEDCs 33

Contemporary urban growth in MEDCs 36

Urban land use and population patterns in cities 40

■ ■ ■

Questions and Answers
About this section 46

Q1 Population (I) 47

Q2 Population (II) 52

Q3 Rural settlement (I) 56

Q4 Rural settlement (II) 61

Q5 Urban settlement (I) 65

Q6 Urban settlement (II) 70

Introduction

About this guide

This guide has been written to help you prepare for OCR Specification A AS Geography Unit 2, which examines the content of **Module 2681: The Human Environment**. The book contains three parts:
- an Introduction, which explains the assessment structure and reveals the techniques for answering structured and extended-answer questions
- a Content Guidance section, which includes matrices outlining the specification's content — key themes and questions that are likely to form the substance of Unit 2 are examined
- a Question and Answer section, which includes six specimen questions, each of which is accompanied by two sample student answers, ranging from grade-E to grade-A standard, interspersed by examiner's comments

The aim of Module 2681

The Human Environment is one of three modules that make up the AS specification. It is worth 90 uniform marks (i.e. 30%) of the 300 marks that make up the whole examination.

Module number	Module name	Unit test length	Raw marks	Uniform marks	AS weighting (%)
2680	The Physical Environment	$1\frac{1}{4}$ hours	100	120	40
2681	The Human Environment	1 hour	75	90	30
2682	Geographical Investigation	$\frac{3}{4}$ hour	75	90	30

AS Geography (OCR A): scheme of assessment (from June 2003)

The specification content of Module 2681 covers population and settlement topics:
- population distribution; change; migration
- rural settlements in MEDCs
- urban settlements in MEDCs
- urban settlements in LEDCs

These topics are studied at a range of scales, from local (e.g. settlement site) to global (e.g. population change). The interaction between people and their physical and human environments, whether intentional or unintentional, is an important theme. Some topics, such as national population distribution, rural population change in an MEDC and housing problems in an LEDC urban area, require you to study specific places and examples.

Examination skills

This guide emphasises the importance of examination technique at AS. It assumes that you already have a sound knowledge and understanding of the specification content. You should remember that success at AS requires not just sound knowledge and understanding but also effective technique. Sensible revision is not an indiscriminate learning of content. Topics should be revised within the context of the key themes and questions through which they are likely to be assessed. If you have prepared thoroughly, there should be few, if any, questions in the examination that explore themes not already familiar to you.

Poor examination technique is most apparent when candidates fail to apply knowledge and understanding appropriately to a question. This type of error is all too common and is a major reason for loss of marks at AS. Of course, some marks are available simply for showing knowledge and understanding of a topic. However, to achieve a high standard in the more demanding questions, candidates must apply this knowledge and understanding appropriately.

The nature of structured questions

The Human Environment unit test is a 1 hour written paper comprising three structured questions. One structured question (containing several parts) is set on population and two questions are on settlement, although there can be some overlap. Structured questions require short answers of between two and ten lines. Each question is built around stimulus materials such as sketch maps, charts, diagrams and photographs. The questions are designed to test the following areas:
- your knowledge and understanding of the specification content
- your ability to apply knowledge and understanding in unfamiliar contexts
- your competence in using geographical skills

The demands of structured questions

Structured questions have a gradient of difficulty. The initial questions require mainly descriptions and definitions and are less demanding than the later ones, which are based on explanations and understanding.

Stimulus materials such as maps and charts do not, in themselves, provide answers to questions on the examination paper. Rather they act as a catalyst, inviting you to show wider knowledge and understanding and consider related aspects of the topic. For example, a table of data might show how birth and death rates vary amongst a selection of countries. However, the questions will not ask you to plot the data as a chart or graph, or to make statistical calculations to find the mean birth or death rate. Instead, the sorts of question you might expect include:
- give a definition of the term *natural increase*
- give a summary description of the pattern of population change

- state and explain the factors that influence variations in death rates
- state and explain how age structure influences birth rates

The marks for each question are given in the right-hand margin of the examination paper. These, together with the number of lines, indicate the length of answer required.

How to answer structured questions

- Study the stimulus material carefully and read through all parts of the question before attempting to answer it. To avoid repetition in your answers, you must look ahead to see how the topic is developed.
- Make sure that you understand precisely what each question is asking you to do.
- Take special care with the command words and phrases in each question (see below).
- Plan your answers before writing. You have a limited space in which to write, so you cannot afford to make mistakes or write carelessly. Your answers should be precise and to the point.
- The lines provided for each answer are a guide to a suitable length for your answers. Don't write overlong answers. These waste valuable time and are therefore self-penalising, they produce untidy looking scripts, and they defeat part of the purpose of structured questions, which is to generate answers that are accurate, precise and economical.
- Where appropriate, include examples. In short-answer questions these can be a simple 'e.g.' If your answer is marginal according to the examiner's mark scheme, an appropriate example might just tip the balance in your favour.

Command terms used in AS geography examinations

All examination questions contain command words and phrases which tell you what you must do. You must adhere *strictly* to the instructions of these key words and phrases. In particular, you must distinguish between commands that ask for description and those that require explanation.

Describe.../Define.../What is meant by...?/Name.../State...

These are common description commands. They require knowledge rather than understanding. An accurate word picture, expressed in precise language, is sufficient. Descriptive tasks are less demanding than those that require explanation, and are usually found in the initial sub-questions.

Explain.../Why...?/Give reasons...

These are the most common explanation commands. They usually require some understanding of causes, processes and outcomes, as well as knowledge. You should note that some questions will ask you both to describe and to explain. In your answer, the description and explanation should be quite separate, and this distinction should be clear to the examiner.

Extended-answer questions

The specification states that you must study actual examples, including the population geography of:
- a more economically developed country (MEDC) and a less economically developed country (LEDC)
- a rural region in an MEDC which has undergone significant population and socio-economic change in the last 30–40 years
- an urban settlement in an MEDC where recent economic or social or demographic change has caused environmental problems
- an urban settlement in an LEDC where urbanisation has caused environmental problems

At least one sub-question on each examination paper will require more extended writing (e.g. a side of A4). This question is likely to provide an opportunity to write about a specific place or area you have studied. A typical question might have the following form:

With reference to a named rural region in an MEDC, show how recent changes in its service provision have affected some inhabitants more than others. (10 marks)

Answers that fail to refer in some detail to a specific example are unlikely to achieve more than half the available marks. The temptation to write generalised answers (even if they are accurate) in response to this type of question must be avoided.

Content Guidance

This section provides an overview of the key terms and concepts covered in **Module 2681: The Human Environment**.

The content of Module 2681 falls into two main areas:
(1) population: pattern, process and change
(2) rural and urban settlement: pattern, process and change

You require good knowledge and understanding of the specification content if you are to do well at AS. Doubtless, much of your revision will concentrate on this area. However, when you revise you must always consider how the examiners might test your knowledge and understanding. That is why, alongside the specification content for each environmental system, this guide provides a list of key questions together with some amplification. It will pay you to study these key questions carefully. Many of the themes they cover are likely to be examined. Revising and preparing answers to them will give you a head start in the AS examination.

Population

Population distribution

Key questions

Topic	Detail	Key questions
Population distribution	The factors influencing population distribution: physical, economic, social, political	• What are the main physical influences on population distribution? How do these influences operate? • What are the main economic influences on population distribution? How do these influences operate? • What are the main social influences on population distribution? How do these influences operate? • What are the main political influences on population distribution? How do these influences operate?
	Spatial variations in birth and death rates	• How are birth and death rates measured? • What are the main differences in birth and death rates at a global scale? • Why do birth and death rates vary at a global scale?
	Spatial variations in age and sex structures	• What is meant by age and sex structure? • What are the main differences in age and sex structures at a global scale?

Key questions answered

What are the main physical influences on population distribution?
The principal physical influences are relief (altitude, slope angle and aspect), climate (precipitation, length of growing season), soil quality, geology and disease.

How do these influences operate?
The physical environment offers both significant limits and opportunities for human settlement. Although population is repelled by extreme conditions, such as very low

temperatures or steeply sloping areas, people are attracted to areas of more moderate conditions. Continental interiors with their extreme climates are sparsely populated whereas areas closer to the coast are more densely settled. Rivers provide an example of the negative and positive influences of an individual factor. People are attracted to rivers for water supply, food supply, fertile soils and transport; on the other hand, the risk of flooding or water-borne diseases can have a discouraging influence.

In different parts of the world, the same factor can have different influences. In the mid to high latitudes, such as northwestern Europe and New Zealand, population density decreases with increasing altitude. In the tropics, such as eastern Africa, a gain in height can moderate the climate and so areas of higher altitude generate higher population densities than lower zones.

What are the main economic influences on population distribution?

The principal economic influences are agricultural productivity and the location of economic activity such as raw material extraction, manufacturing industry and service activity.

How do these influences operate?

Regions that have low agricultural productivity, such as arid and semi-arid areas, support fewer people than areas where crops and livestock can be produced in greater quantities and on a regular basis. Traditionally, the fertile valleys of large rivers such as the Ganges, Nile and Danube have offered good opportunities for settlement. The availability of minerals such as fuels (coal, oil) and ores (iron, copper) may encourage settlement in areas with such resources, which can be the basis of substantial industries. Not all such resources attract significant numbers of people, however, as other factors act as negative forces. The oil in northern Alaska and iron ore in western Sahara have not attracted large numbers. Those regions where service activity has become important, such as western Europe, continue to support high population densities. Here, local resources are no longer as crucial as they used to be, as many materials are traded with the money generated by service activity.

What are the main social influences on population distribution?

There are two principal social influences: migrations and culture.

How do these influences operate?

There have been substantial movements of people at a variety of scales. For example, many millions of Europeans moved to North America and so influenced the population distributions both of the places they left and of the continent they moved to. Movements within a country can affect population distribution, as in the UK where there has been a general urban to rural shift over recent decades. People may be influenced by cultural factors such as religion: when India was partitioned, large numbers of Hindus and Muslims moved to the country (India and Pakistan) where their religion predominated.

What are the main political influences on population distribution?

Political influences that affect population distribution include government plans and programmes.

How do these influences operate?
Governments can influence directly or indirectly where people live. In Indonesia, there have been attempts to move people away from the crowded island of Java to more remote islands. The relocation of a capital city can encourage people to move to the new capital and so alter population distribution, as has happened in Nigeria and Brazil. The New Town policy in the UK encouraged people away from conurbations towards settlements such as Telford and Harlow.

How are birth and death rates measured?
The most common measures are crude birth and death rates, CBR and CDR. These rates are ratios of the number of live births or deaths to the total population, usually expressed per 1000 people. These are simple measures, but have their drawbacks. CBR uses the entire population in its calculation, when, in reality, no males and a significant number of females (those too young or too old) are unable to give birth. There are several alternative measures: age-specific rates relate births/deaths to particular age groups, while the total fertility rate (TFR) is the average number of live births per woman during her lifetime. In the UK, the TFR is around 1.7, while in some LEDCs it can rise to 6.5.

What are the main differences in birth and death rates at a global scale?
Over the past 200 years, and in particular since 1950, there has been unprecedented growth in the world's population. From just over 1 billion in 1800, past 2.6 billion in 1950 to just over 6 billion in 2000, global population change has been the result of a substantially higher number of births than deaths.

During the past decade and a half, some significant differences in the global pattern of population change have emerged. The MEDCs tend to experience low growth rates — the result of birth and death rates being at relatively low levels. By 2050, the MEDC share of global population is likely to fall to about 12%, from 20% in 1995.

In most LEDCs, population growth remains rapid as birth rates have remained at relatively high levels while death rates have fallen.

There is significant diversity within these broad trends. North America, Oceania and Europe are dominated by MEDCs, although it is only in Europe that birth and death rates are almost the same. The continents of Africa, South America and Asia contain most of the LEDCs but Africa has by far the highest birth rate and the highest death rate.

Such average figures have a value in outlining the general picture, but individual countries or regions within a country can have quite different patterns of birth and death rates.

Continent	Birth rate per 1000 (2001)	Death rate per 1000 (2001)
Africa	40	14
America, North	14	9
America, South	25	7
Asia	24	8
Europe	12	11
Oceania	19	9

Why do birth and death rates vary at a global scale?

Several factors influence the level of birth and death rates. The economic status, both of the individuals concerned and of the country where they live, is one factor: generally, as economic security and status increase, CBRs and CDRs decrease. Money is available to pay for a good standard of health care, which in turn results in a lower maternal and infant mortality. Fewer births are needed to ensure that at least some children survive into adulthood. Lower economic status in LEDCs is often associated with subsistence agriculture or informal service employment in which children can be economic assets. In MEDCs, the economic costs of bringing up children are substantial and may encourage people to limit their family size.

The role and status of women is an important influence. Many women in LEDCs have little formal education and tend to marry at a young age, which means they start having children early and often continue to be occupied in child bearing and rearing. In MEDCs, the situation is reversed, with women achieving higher educational levels, delaying the age of marriage, and pursuing formal careers.

The level of health care directly affects both birth and death rates. The availability of family planning such as contraception makes it practical for couples to limit the size of their families. The prevention and treatment of disease can reduce mortality. Standards of health care are significantly higher in MEDCs than in LEDCs. The provision of clean water has a major influence on death rates.

In LEDCs, many people have inadequate diets and so are less resistant to disease. In MEDCs, diet can also influence death rates, for example through heart disease and tobacco and alcohol consumption.

Birth rates are influenced by religion and culture. The use of family planning is discouraged by some religions. In some cultures, having many children is seen as a sign of tribal or family strength or an individual's virility.

What is meant by age and sex structure?

This is the make-up of a population according to age groups and sex. Normally, 5-year age groups are used for each gender. These are represented on a type of bar chart known as a population pyramid. Various ways of measuring the structure can be used. Most common is the dependency ratio, calculated by dividing the number of children plus elderly by the number of adults.

What are the main differences in age and sex structures at a global scale?

The principal difference is the relatively even distribution of both age and sex up to the age of 60 in the MEDCs and the more youthful populations found in the LEDCs. The MEDCs are showing signs of an ageing population and, among the elderly, there are more females than males. Many LEDCs have a tapering shape to their pyramid, but with falling birth rates the pyramid base is starting to narrow. As mortality (especially infant mortality) declines, LEDC pyramids are beginning to show signs of straightening in their younger age bands.

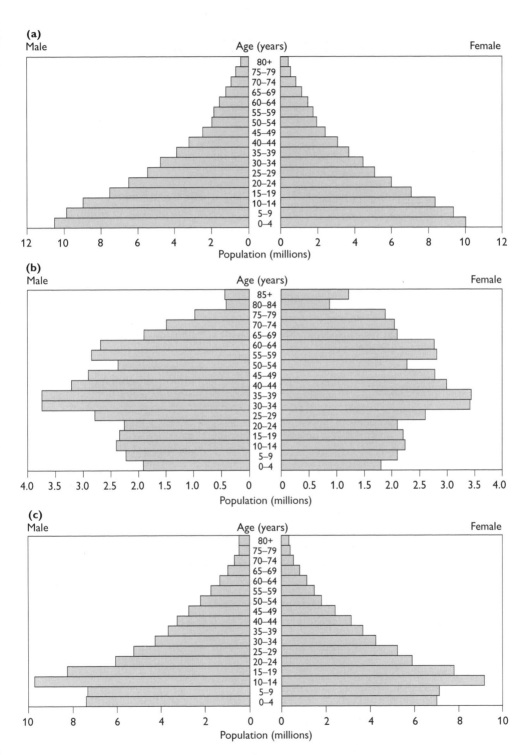

Population pyramids: (a) Pakistan, (b) Germany and (c) Bangladesh, 2001

Population change through time

Key questions

Topic	Detail	Key questions
Population change through time	The factors influencing population change: fertility, mortality, migration, age–sex structures	• What is meant by population change? • How are rates of population change calculated? • How and why do fertility and mortality change through time? • How does migration influence population change? • How do differences in age–sex structures influence population change?
	The demographic transition and its value in understanding population change through time	• What is the demographic transition? • How valuable is it in understanding population change through time?
	The ways in which governments can influence population change	• What are the ways by which governments can influence population change?

Key questions answered

What is meant by population change?

At a global scale, the balance between the number of births and the number of deaths is responsible for population change. When the number of births exceeds the number of deaths, **natural increase** results; **natural decrease** occurs when the number of deaths exceeds the number of births.

How are rates of population change calculated?

The % natural increase or decrease per year is calculated using the formula:

$$\frac{CBR/1000 - CDR/1000}{10}$$

In 2002, the % global population change was $\frac{25-9}{10} = 1.6\%$.

How and why do fertility and mortality change through time?

Economic and social changes within MEDCs over the past 200 years have led to a decline in fertility and mortality. Economic growth led to rising standards of living that allowed people to enjoy better diets and housing conditions. Health care improved, as did educational levels, which then led to changes in attitudes among people as regards family size. The change from a largely rural society to one where the vast majority of people lived in towns and cities also influenced people to change their attitudes towards family size.

How does migration influence population change?

Migration is the permanent or semi-permanent change of residence by an individual or group of people. Many migrations are selective, that is, some groups of people are more likely to move than others. If an area receives large numbers of retired migrants, then natural increase in that area may reduce. An area might undergo significant out-migration of young people and this would reduce its rate of natural increase. Some locations receive large numbers of young people, so the rate of natural increase might be high and could lead to population growth.

How do differences in age–sex structures influence population change?

The younger the age profile of an area, the lower its death rates will be. Many LEDCs have a high proportion of their population as young adults or children, so death rates are as low as, if not lower than, in many MEDCs. This is particularly the case in those countries undergoing rapid economic growth, sometimes called newly industrialising countries, such as Taiwan and South Korea. In the MEDCs, the high proportion of middle-aged and elderly people results in higher death rates — above the level expected in some of the more developed LEDCs. Some MEDCs have experienced periods of natural decrease when deaths were greater than births; by contrast, the LEDCs are experiencing natural increase. As there are so many children in LEDCs, the next few decades are likely to continue to see significant natural increase in these countries, as these children become young adults, marry and have families of their own; this is known as demographic momentum.

What is the demographic transition?

This model describes the change from conditions of high fertility and mortality to low fertility and mortality. In its basic form, four stages are identified:

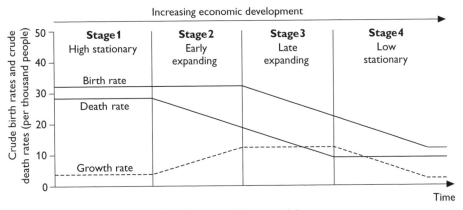

The demographic transition

- Stage 1: both fertility and mortality are high and as a result population growth is low.
- Stage 2: mortality declines while fertility remains high, resulting in rapid population growth.
- Stage 3: mortality continues to fall while fertility begins to decline, resulting in the rate of population increase slowing.
- Stage 4: by this stage, both fertility and mortality are at low levels so that population growth is either very low or not happening.

How valuable is it in understanding population change through time?

It is based on the experience of some parts of western Europe between the mid-eighteenth and mid-twentieth centuries. The economic and social conditions were such that there was economic growth and many people enjoyed rising standards of living; medical care improved; and society transformed from rural to urban so that factors such as education, the role and status of women and democratic political systems improved.

It is unwise to use this descriptive model to predict population change in other parts of the world. Both population growth rates and absolute numbers of population increase were lower in western Europe than in LEDCs in the last 60 years. The rate at which mortality declined in Europe was relatively slow, whereas in LEDCs it has been rapid, largely due to the transfer of medical technology from MEDCs rather than to an improvement in living standards.

In some LEDCs, urbanisation has been a slow process, as has the move away from attitudes that encourage high fertility levels, such as the perceived need for large numbers of children due to previously high rates of infant mortality. The continuing economic advantage of children through their employment also maintains high fertility levels.

There are significant cultural differences between most LEDCs and countries in western Europe, for example in the varying emphasis that different religions place on the role of the family. Increasingly during the twentieth century, government influence extended to include population matters — a factor not present for the first three stages of the model in western Europe.

What are the ways by which governments can influence population change?

Governments of all types have been increasingly involved in population policies. Encouraging a fall in mortality has been commonplace, from the measures in nineteenth-century Europe to improve water supply, sewage disposal and housing conditions, through to the mass inoculations against diseases such as smallpox and polio in both MEDCs and LEDCs in the twentieth century.

Influencing the level of fertility has also been a strong theme for all types of government. Some countries have pursued strong birth control programmes (anti-natalist) while others have encouraged a rise in birth rates (pro-natalist). These measures can be voluntary, such as the use of advertising and education, or they can be imposed, such as limiting the number of children a couple can have.

Population change through space

Key questions

Topic	Detail	Key questions
Population change through space	Types of migration	• What is migration? • What are the main types of migration?
	Causes of migration	• Why are some people more likely to migrate than others? • How does perception influence migration? • What are the sorts of obstacle that influence migration?
	Consequences of migration	• How does migration affect the source area of the migrants? • How does migration affect the destination of the migrants?

Key questions answered

What is migration?
Migration is a permanent or semi-permanent change of residence. Net migration is the difference between the number of migrants entering an area and the number of people leaving. When more people move into an area than leave it, there is net migrational gain; when more leave than move in, net migrational loss is recorded.

What are the main types of migration?
It is possible to have a very complicated classification of migrations, but there are some basic types. The divide between rural and urban areas results in four types of migration: rural–urban, rural–rural, urban–rural and urban–urban. In LEDCs, rural–urban migration results in net migrational gain for urban areas, whereas in MEDCs, most migrations are either urban–urban or urban–rural, the latter being important in the process of counter-urbanisation.

It is also possible to classify migrations by how the migrants move. For example, step-wise migration describes the movement up a settlement hierarchy from village to town to city. Chain migration is strongly influenced by family links, for example movement between a rural region and a particular city.

The scale at which migration occurs varies from the local, such as within a town, through the regional, such as from one part of a country to another, to international, when migrants cross national borders.

Why are some people more likely to migrate than others?

Migration is a selective process influenced by factors such as age, gender, stage in family life cycle, education and employment. Young adults are often the most mobile, as they are at the start of their independent lives and do not yet have to consider factors such as the effect on children of a move. In later years, retired people have relative freedom in where they live and migration flows of the over-60s have become an important part of the population geography of MEDCs. Young men are common among the rural–urban migrants in LEDCs, as the employment possibilities in the cities are seen to be mostly for males, such as in factories; wives, at first, may stay in the village until their husbands have more secure jobs. Staying longer in education raises the individuals' skills levels and can lead to them being more aware of opportunities and so more likely to migrate.

How does perception influence migration?

Flows of migrants are determined by thousands of personal decisions made by people assessing whether it is in their best interests to move or stay. Such decisions are strongly influenced by the perceptions — the subjective views — of the individual. Individuals have personal opinions about their current situation, for example their employment prospects or their present housing. Balanced against these are their mental images of possible destinations derived from various sources, such as the media or relatives and friends. Both the existing location of the potential migrant and the destination will have advantages and disadvantages, so each person weighs up the balance of whether to move or stay. When the positives of the destination predominate, then a migration often takes place.

What are the sorts of obstacle that influence migration?

Between the existing home of the migrant and the potential destination, a variety of obstacles might exist. These can be physical (e.g. a mountain range, a sea or poor transport links), political (e.g. as international borders that have strict immigration controls), or legal (e.g. the need for work permits). Sometimes migrants encounter an intervening opportunity before reaching their intended destination. They may find good employment in a town along the route. Many migrants end up staying in the port they land at when entering a new country as they discover opportunities there, rather than moving further into the country.

How does migration affect the source area of the migrants?

As migration is selective, the source area has fewer of certain groups in its population. In rural areas in LEDCs — parts of Africa in particular — there may be fewer young adult males as a result of their migration to urban centres in search of employment. In remote rural regions in MEDCs, such as upland areas like mid and north Wales or the Massif Central, France, there is a general absence of both males and females amongst the younger age groups as a result of out-migration in search of employment and social opportunities.

How does migration affect the destination of the migrants?

Destinations will also show evidence of the selective nature of migration. Regions receiving flows of retirement migrants have a top-heavy age–sex pyramid, with a greater proportion of females among the more elderly.

In both source and destination areas, there are social, economic and political consequences of whatever migration flows the area is experiencing. Gaining or losing certain population groups will affect the type of goods and services that can be offered. Contact with a different way of life, such as the contrast between rural and urban areas in the LEDCs, can expose people to new ideas and change the way they think about issues such as democracy and the role and status of women.

Rural settlement
Rural settlements in MEDCs
Key questions

Topic	Detail	Key questions
The location of settlements	Settlement site and situation	• What are the main site factors? • How does situation influence a settlement's location?
	Settlement form (shape)	• How and why do settlement forms vary?
Settlement patterns	Types of settlement pattern	• What are nucleated and dispersed rural settlement patterns? • How do nucleated and dispersed patterns develop?
	The evolution of settlement patterns	• What is meant by the evolution of settlement patterns? • What are the typical main stages of settlement evolution?
Settlements as central places	Central places, range, threshold, hierarchy and centrality	• How do settlements act as central places? • What is the difference between range and threshold? • Is there any organised pattern to the availability of goods and services in a rural region? • Why are there exceptions to the ideal patterns? • What aspects of a central place determine its centrality? • What aspects of the surrounding area determine the size and shape of the market area? • What is a settlement hierarchy? • What factors are responsible for a change in a settlement's position in a hierarchy?

OCR (A) Unit 2

Key questions answered

What are the main site factors?

Site is the actual land a settlement is built on. To understand how site factors influenced where a village was built, it is important to appreciate that the first settlers possessed limited technology with which to alter the natural environment. A reliable fresh-water supply was important, but not at the risk of flooding. Gentle slopes and shelter from strong prevailing winds are other factors that tended to be considered. In areas with permeable surface geology, such as chalk, wet-point sites where springs emerge tended to attract settlement. This contrasts with areas subject to flooding, such as areas of clay where dry-point sites were at a premium.

How does situation influence a settlement's location?

Situation is the location of a settlement in relation to resources outside its site. At the local scale, situation would refer to factors such as where woodland, arable and pastureland might be. At the larger scale, estuaries, lowest bridging points, gaps through upland regions and the location of other settlements are examples of situation factors.

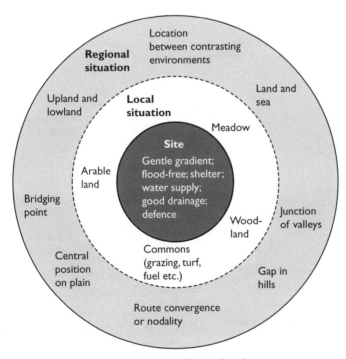

Site, situation and village development

How and why do settlement forms vary?

Form is the shape of the settlement — how the buildings, roads and open spaces are arranged. Linear villages are spread out along a road, while the clustering of houses and farms around a central village green is another common shape.

A range of physical and human factors influence village form: the nature of the site, for example a village spread out along a river terrace above the floodplain; the line of roads and the presence of junctions; how the society that founded the original settlement was organised; and more recent planning decisions, for example permission to build a small housing estate on the edge of a village.

What are nucleated and dispersed rural settlement patterns?

The grouping together of farms and houses in clusters (villages and hamlets) is known as nucleation. Where the farms tend to be scattered across an area as isolated buildings, this is known as dispersion. The settlement pattern in a rural region is the overall distribution of isolated farms, hamlets and villages. When clusters dominate, the pattern can be described as a nucleated one; a dispersed pattern is where isolated farms dominate. Most rural patterns are made up of a combination of isolated farms, hamlets and villages.

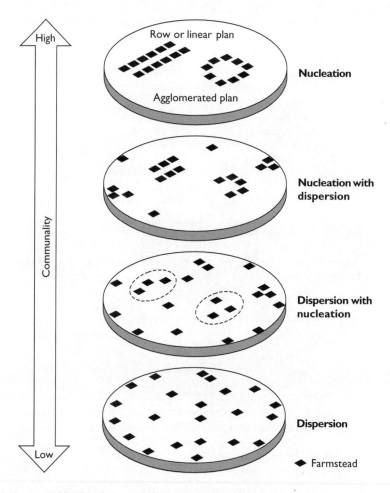

Nucleation and dispersion in settlement patterns

How do nucleated and dispersed patterns develop?

In Britain, a basic division can be recognised between the dispersed pattern in upland areas and the nucleated pattern of the lowlands. The factors influencing this contrast tend to be both physical and cultural.

In upland regions, the physical resources for agriculture are generally poor; temperatures are lower, rainfall is higher, soils are thinner, more acidic and often poorly drained, and the growing season is shorter than in lowland areas. Agriculture in the uplands, of necessity, needed to be low-intensity, with individual farms covering large areas. The farmhouse and associated buildings located centrally in its land, so settlement dispersal was the result. In lowland areas, the generally favourable conditions allowed more intensive farming, often with a strong arable element, operating in smaller farms and encouraging clustering in villages.

In upland regions, cultural factors also encouraged dispersal. Here, people tended to have more individual freedom and farmers often owned their land and preferred to live on it. In lowland regions, there was a tradition of feudalism, which meant that land was owned by a small elite and the majority of the people were given land in return for various services on the lord's estate. Land was often worked cooperatively. These factors encouraged nucleation in villages.

It is important to appreciate that when describing the settlement pattern of an area, one should assess the dominant style, either dispersed or nucleated. Upland areas have some nucleation and lowland regions have dispersed farms.

What is meant by the evolution of settlement patterns?

Evolution refers to change through time, so with regard to settlement patterns it is the change in locations, numbers and types of settlement through time.

What are the typical main stages of settlement evolution?

A number of stages in the settling of Britain can be recognised on an Ordnance Survey map. The earliest tends to be the presence of earthworks (tumuli) that can vary in age from Neolithic, through Bronze Age to Iron Age.

Evidence of the Roman period comes from elements of place names, such as 'cester' (Cirencester), 'caster' (Doncaster) and 'chester' (Colchester).

Between the end of Roman rule and the arrival of the Normans, the invasion and colonisation of Britain by the Anglo-Saxons and Scandinavians had a major influence on settlement patterns. Evidence again comes from parts of place names, such as 'ing', 'ham', 'ton', 'ford', 'weald' and 'den' from the Anglo-Saxons, and 'by', 'garth' and 'thorpe' from the Scandinavians. The Anglo-Saxons were mainly in the Midlands and south of England with Scandinavian influence strongest in the north and east of England. By the time of Domesday in 1086, most existing settlements in lowland England had been established.

The Norman stage and, later, the medieval period saw the establishment of religious settlements such as abbeys and priories. Halls, manors and granges are evidence of this period and are often seen in place names. The existence of a castle would also be an element of the pattern.

In the nineteenth century, enclosure often resulted in isolated farms being established in lowland areas, and the increased ability to drain land allowed farms to move into low-lying areas such as the Fens and Somerset Levels. Canals and railways were capable of encouraging the growth of settlements they passed through.

In the twentieth century, New Towns were built to try to relieve pressure in some of the conurbations, for example Telford (West Midlands) and Harlow and Stevenage (London).

How do settlements act as central places?
Central places are settlements that provide goods and services to their own population and the people of the surrounding area — the market or catchment area. Different types of goods and services are found in settlements depending on various factors, such as the population of the settlement, the population of the area surrounding it and the settlement's accessibility.

What is the difference between range and threshold?
Range is the distance that people travel to obtain a particular good or service. Threshold is the number of people needed to support a good or service to keep the business viable.

Threshold is not just about numbers of people. Factors such as age structure and, in particular, average income will influence the types of goods and services available in a central place. A settlement whose inhabitants have above-average incomes will be able to support more medium- and higher-order functions, such as restaurants, than a settlement with a similar total population but where the incomes are lower.

Is there any organised pattern to the availability of goods and services in a rural region?
Some goods and services are required infrequently by any one individual (high-order functions), while others are in demand most weeks (low-order functions). It does not make economic sense, therefore, for all goods and services to be available everywhere, as the higher-order functions need a greater threshold. In addition, people are prepared to travel further to obtain the higher-order functions. The ideal distribution of central places provides a population with a complete range of goods and services with the minimum of travel. Villages have a small number of functions; market towns have the same functions but more of each type as well as a few extra functions not found in the villages; a city has all the functions found in the smaller settlements surrounding it, and lots of each type, as well as the functions that mark it out as a more important settlement in functional terms. For example, there could be a playgroup and a primary school in a village; the market town might have several playgroups, three or four primary and junior schools and a secondary school; the city will have many playgroups, primary and junior schools, several secondary schools, a college and a university.

Why are there exceptions to the ideal pattern?

As personal mobility has increased, especially since 1950, people have been able to travel further than their local central place. Villages are no longer the agricultural settlements they used to be now that commuters are working and obtaining their goods and services outside the village or even the local market town; people often shop where they work. The development of superstores has led to the closure of many small retail outlets.

The provision of both education and health care has been rationalised, involving the closure of smaller units and the concentration of facilities into fewer but larger units.

Some settlements do not act solely as central places. Tourist resorts, for example, provide goods and services for all their visitors and so possess more functions than their population alone might indicate; this is called 'over-serving'. A dormitory or commuter village may have fewer functions than its population might indicate, as its residents tend to obtain their goods and services from the metropolitan centre where they work; this is called 'under-serving'.

What aspects of a central place determine its centrality?

The centrality of a settlement is a measure of how important the settlement is in terms of providing goods and services. It can be measured in various ways: the total numbers of goods and services, the variety of functions and the area of retail floorspace are direct measures of centrality. Other measures are indirectly related to providing goods and services, but still offer an insight into centrality, such as car parking spaces and a settlement's accessibility as shown by bus and/or rail routes. Perhaps the simplest indicator is total population and this is closely related to threshold.

What aspects of the surrounding area determine the size and shape of the market area?

The greater the centrality of a settlement, the larger is the size of market area. A city attracts people from a much larger area than a market town. If the population density of a region is low, then market areas tend to be larger, as it takes a greater area to achieve the threshold required to support the functions of the central place. Theoretically, the shape of market areas should be hexagonal, as this shape fits together, avoiding overlap or gaps between market areas. Factors such as physical features may alter this pattern. Market areas tend to extend further across upland areas, as population density is lower and so a larger market area is required to reach the threshold. The location of competing central places will affect the size and shape of a market area. Market areas can be extended in the direction of a transport route, such as a major road.

What is a settlement hierarchy?

A settlement hierarchy is the arrangement of the settlements in an area into an ordered system depending on their relative status as regards supplying goods and services. A hierarchy ensures that low-order functions are available in most settlements but that functions with a high threshold and long range are found in just a few settlements. At the bottom of a settlement hierarchy are the hamlets and villages, just above these are the market towns, with larger towns and cities occupying higher levels in the arrangement. At each step up in the hierarchy, there are fewer of that type of settlement and therefore they tend to be spaced further apart; there are lots of villages but few cities.

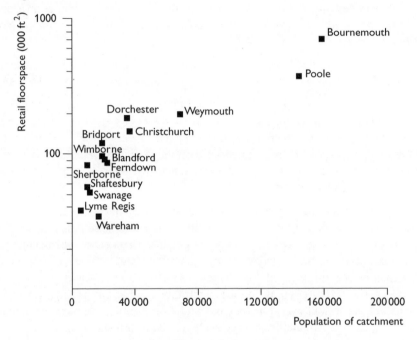

Hierarchy of central places in Dorset

What factors are responsible for a change in a settlement's position in a hierarchy?

Over a time scale of 50 years or more, an individual settlement's population can either grow or decline and so affect the threshold and thereby its position in the hierarchy. In the mid-eighteenth century, settlements such as Birmingham, Bradford and Brighton were towards the bottom of the hierarchy in their areas. They grew as manufacturing industry or tourism gave their economy a boost and so they became higher-order central places.

On a local scale and often over a shorter time period, a settlement can rise up or fall down the local hierarchy. The construction of an improved transport link such as a motorway might improve a settlement's accessibility and it may grow. A small village can become a busy commuter village, a market town may become a major distribution centre. The closure of a local mineral working such as a coal mine or stone quarry may lead to a reduction in employment, out-migration and a loss of threshold. Government planning in the form of a key settlement policy may support one settlement at the expense of others as services are rationalised and concentrated in the chosen village.

Population change in MEDCs since 1960

Key questions

Topic	Detail	Key questions
Changes due to migration	Rural–urban migration and rural depopulation	• What is rural depopulation? • Why do people leave rural areas?
	Urban–rural migration and counterurbanisation	• What is counterurbanisation? • Why do people move to rural areas?
	The impact of migration	• How does migration affect age structure? • What is the impact of migration on different socio-economic groups? • How and why have rural services been affected by migration?
	The contrasting experience of rural areas close to, and remote from, large urban areas	• What has been the difference between rural areas in different locations?

Key questions answered

What is rural depopulation?

This is the absolute decrease in the population of a rural region. In MEDCs, it has mostly been brought about by net out-migration, although with increasingly elderly populations in some rural areas, natural decrease might also be a factor. Some rural regions such as the northwest Highlands and Islands of Scotland and parts of the Massif Central in France have suffered depopulation for decades.

Why do people leave rural areas?

The loss of employment opportunities is perhaps the main factor in out-migration. Agriculture, both in lowland and highland regions, has seen a sustained reduction in its demand for labour since the 1960s, and other sectors such as quarrying and forestry have also seen numerical decline in workforces. In some areas, second home ownership can raise house prices beyond the ability of local people to afford them and so young people in particular leave. There is also the perception among some of a more fulfilling lifestyle in metropolitan centres with their entertainment facilities and faster pace of life.

What is counterurbanisation?

This is a shift in population from the larger urban settlements to rural areas, including small market towns, villages and hamlets. It was first identified in the USA and became a strong process in western Europe during the 1970s and 1980s.

Why do people move to rural areas?

There has been a decentralisation of employment away from urban settlements as businesses take advantage of factors such as cheaper land, more space, lower taxes and less trade-union influence. For individual households, rural areas offer an improved physical environment in which to live. For those in work, the increased commuting to employment and services in urban areas is offset by the increased personal mobility that many enjoy. This has come about due to increased levels of car ownership and improvements in transport infrastructure, such as electrification of railways and improved roads. Some people are able to conduct increasing amounts of their work from home via e-mail, fax and the internet, so that a high-cost central location in a city is redundant. In an ageing population, for the greater percentage of retired, healthy and relatively affluent people, the smaller towns and villages offer a better quality of life, usually at a lower cost, than the larger urban settlements.

Areas of high environmental quality, such as coastal areas and national parks, are very popular for retirement migration.

How does migration affect age structure?

In regions undergoing depopulation, the out-migrants tend to be from the younger sections of the population. The population pyramid tends to have a narrow base and middle as the child-rearing adults and their young families move away, leaving a relatively high proportion of elderly people.

Where a rural region is receiving people from the process of counterurbanisation, the age structure tends to be something of a mirror image to the depopulated area. Families and their young children move out of cities, swelling the under-15 and 30–late-40s age groups in the rural region.

Rural districts receiving retirement migration show a top-heavy age structure.

What is the impact of migration on different socioeconomic groups?

Both retirement migration and the movement of families away from cities tend to be more common among the higher socioeconomic groups. Those leaving rural areas are generally from the younger age groups — both those who continued their education beyond school and those who did not. Those who came from the agricultural sector may live in rented accommodation associated with their work and may find it hard to afford the higher housing costs usually charged in cities. Perhaps the greatest impact is that on service provision.

How and why have rural services been affected by migration?

Small independent retailers in rural areas have been closing in large numbers. The increase in personal mobility has allowed rural dwellers to shop in the supermarkets and superstores located in nearby towns and cities. The companies that run such units can take advantage of economies of scale and so sell their products at prices that undercut the independent shopkeepers running village stores. The widespread ownership of fridges and freezers allows people to shop less often but to buy in bulk, which large supermarkets are better able to supply.

In the early 1960s, most villages had a primary school, but today increasing numbers of villages have no school. The absence of children might be due to depopulation, the out-migration of child-rearing adults, or the in-migration of retired people. Economies of scale also apply to schools so that small village primary and junior schools are expensive to run. They may not be able to offer a full range of curriculum opportunities because they have too few staff. Educational authorities running rural areas have tended to rationalise schools, concentrating on fewer but larger schools and transporting children from villages.

Rural areas used to be well served by transport services, both rail and road. With increased car ownership among some groups, there has been a reduction in public transport in rural areas as demand has fallen. Groups such as the elderly, children and teenagers, and those with lower incomes, are often badly affected by the loss of transport services.

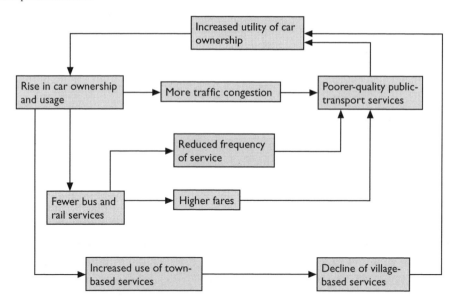

The effect of increasing levels of car ownership on public-transport services

What has been the difference between rural areas in different locations?

Rural areas in remote locations are significantly different in several ways from rural areas close to — that is, within commuting distance of — major urban settlements.

Remote rural areas are often upland areas that continue to depend heavily for employment on agriculture and forestry. Incomes are low and few employment opportunities exist. Net out-migration and depopulation have been occurring for many decades, so that thresholds have declined and services closed. A few places have undergone some revival as people have moved in: these in-migrants have tended to be either retirees or people able to work from home, perhaps by using advances in communication such as e-mail and the internet.

By contrast, rural areas close to urban centres have lost most of their links with agriculture, and have little employment linked to rural activities. Villages have become dormitory settlements for the town or city, with commuters obtaining employment and services in the urban centres. Housing estates, in-filling of plots and conversion of former agricultural buildings such as barns have been aimed at the higher socio-economic groups, where individuals have good personal mobility through high rates of car ownership.

Urban settlement

Contemporary urbanisation in LEDCs

Key questions

Topic	Detail	Key questions
The growth of urban areas in LEDCs	Urbanisation and urban growth	• What is meant by urbanisation and urban growth? • What is the world pattern of urbanisation? • How does natural increase lead to the growth of urban areas? • How does rural–urban migration lead to the growth of urban areas?
	Social and economic advantages for rural–urban migrants	• Why do people leave rural areas? • Why do people move to urban areas?
	Problems of housing, jobs and service provision in LEDC urban areas	• Why is there a shortage of housing in LEDC urban areas? • What are the difficulties in providing jobs and services?

Key questions answered

What is meant by urbanisation and urban growth?
Urbanisation is the process whereby the proportion of people living in towns and cities increases. When a society is undergoing urbanisation there is a relative shift of population from rural to urban places. Urban growth is the increase in the number of people living in towns and cities.

What is the world pattern of urbanisation?
Just about half the world's population lives in urban places, a proportion reached because of high rates of urbanisation over the past 50 years. The more urbanised continents are Europe, North and South America and Oceania, with over 60% of their populations living in urban areas; the less urbanised continents are Asia and Africa with about 30% of their populations living in urban areas.

The process of urbanisation is most rapid in Africa and Asia where, if current trends continue, about half the population will be living in towns and cities in a couple of decades. Within the same time frame, at the global scale, nearly 80% of all urban dwellers will be in the LEDCs. Urbanisation levels in the MEDCs seem to have peaked and counterurbanisation is an established process.

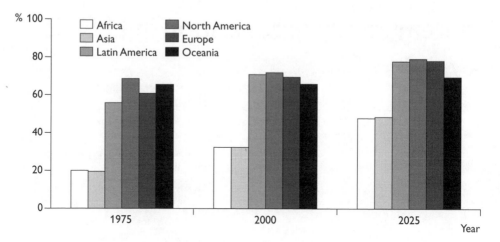

Population in urban areas by continent (%)

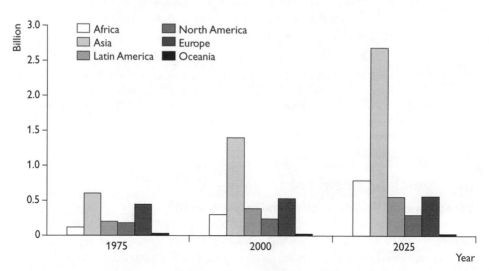

The urban population of the world by continent (billion)

How does natural increase lead to the growth of urban areas?

An excess of births over deaths — natural increase — is common throughout LEDCs. Urban areas are no exception to this trend, with a substantial part of growth, some 60%, coming from this source. Rural areas also experience natural increase, which can lead to rural–urban migration.

How does rural–urban migration lead to the growth of urban areas?

Urban growth not accounted for by natural increase comes from the movement of people from the countryside to urban settlements. Sometimes this movement takes people directly from their village to the urban centre, but very often there is a step-wise movement. This describes the migration of people from their rural settlement up through the various levels of the settlement hierarchy — village, small town, large town, city — until they eventually arrive in a metropolis. There can also be chain migration, which describes the role of family ties between a particular rural region and a city. Once a few migrants from a rural area have made the move to a city and become established, family and friends are often encouraged to follow them, as the earlier migrants provide an initial base in which to stay and search for work and housing. Also, those already in the city will become aware of employment opportunities and pass on this information to their contacts back in the rural area.

Why do people leave rural areas?

Natural disasters, such as floods, droughts and volcanic eruptions, can so disrupt a rural area that people decide to leave. The mechanisation of agriculture reduces job opportunities, particularly amongst labourers hired on a daily basis. Fragmentation of farm holdings into units too small to sustain a family, particularly when natural increase is high, can encourage a move. Rural incomes are generally low, and small farmers and local craftsmen may find themselves in debt, possibly at high rates of interest to local moneylenders. Land, livestock and equipment may be sold to pay the debts, leaving these people little choice but to move to begin their lives again. In some areas, a rigid social hierarchy prevents upward social mobility. Major development projects, such as dams and irrigation schemes, might force the resettlement of people, some of whom prefer to head for the city.

Why do people move to urban areas?

The perceptions many people have of urban life contrast favourably with how they are living in the countryside. Incomes, employment opportunities, education and health care are seen as being better in the cities than in the rural areas, and there is a greater chance of obtaining them in urban areas.

Why is there a shortage of housing in LEDC urban areas?

The lack of housing is most severe for the poorer members of urban society. They cannot afford to spend much on housing from their very limited incomes. On the other hand, the scale of growth in the urban population overwhelms the authorities, as they too have limited resources and are unable to provide enough low-cost housing.

What are the difficulties in providing jobs and services?

The growth in population has outstripped increases in regular employment in the manufacturing and service sectors; these jobs are known as formal employment. Such jobs usually require formal education and skills that are difficult for many people to gain. If people cannot obtain regular jobs, they tend to enter the informal sector. This comprises a wide diversity of small-scale, labour-intensive enterprises, both manufacturing and services, relying on traditional skills and technology. Such employment is casual and often based either at home or on the roadside. Garment making, carpentry, food and drink selling and activities that are criminal, such as drug dealing and prostitution, are examples of activities in the informal sector. The importance of this sector should not be undervalued, as it offers a chance of making a living to thousands of urban dwellers.

Services such as mains water, electricity and paved roads are found only in some areas of LEDC cities. As with housing, the scale of growth of the population is too great for the municipal authorities to provide services, in terms of planning and paying for them. Even where there are private companies supplying infrastructure such as power, they too cannot keep up with the demand. There is also the issue of how people who do not have regular incomes can pay for the services.

Contemporary urban growth in MEDCs

Key questions

Topic	Detail	Key questions
Contemporary urban growth in MEDCs	Causes of suburbanisation and exurban growth	• What is meant by suburbanisation and exurban growth? • When did suburbs develop? • What caused suburbs to develop? • What land uses tend to be found in suburbs and why? • Which groups of people tend to live in suburbs and why?
	Problems created by urban growth in MEDCs	• How does urban growth create problems of urban sprawl? • How does urban growth create problems of air pollution? • How does urban growth create problems of waste disposal? • How does urban growth create problems of water supply?

Key questions answered

What is meant by suburbanisation and exurban growth?

Suburbanisation is the growth of urban areas at their edge so that the continuous built-up area is enlarged. Exurban growth refers to growth that is urban in origin, for example people moving out of a city but continuing to work, shop and obtain services there, but its location is physically separate from the city, perhaps in a nearby small town. The term 'decentralisation' is used to describe the outward movement of people, employment and services from the central and inner parts of a city.

When did suburbs develop?

Suburbanisation is a long-standing feature in MEDCs. It was first recognised in medieval times when poor-quality housing developed outside the congested space within a town's walls. When defence from walls became redundant, towns and cities began to spread outwards. There was rapid suburban development in the second half of the nineteenth century and again in the 1930s. After the Second World War, suburbanisation continued but it has become increasingly affected by planning controls.

What caused suburbs to develop?

Once improvements in transport technology, such as railways, metro systems, trams and horse-drawn buses, became common in the later nineteenth century, more and more people were able to escape the high-density housing areas of the central parts of a city. At the same time, rising incomes allowed people to spend more money on their journey to work in order to benefit from living in the more spacious and less polluted suburbs. The same factors applied both in the 1930s and from the 1950s to today. From the 1950s onwards, more people were able to buy their own homes, particularly from among the houses built relatively cheaply by mass construction companies developing suburban estates.

What land uses tend to be found in suburbs and why?

Residential land use dominates as the suburbs represent an attractive environment for people wanting more space than inner areas can offer. Semi-detached and detached housing are the most popular housing types in suburbs. As there is a high threshold, goods and services are also present. Neighbourhood shopping clusters and local parades of shops are common, as are nursery, primary and secondary schools. The decentralisation of manufacturing industry has led to the location of industrial estates in some suburban areas. The increased amount of space available for development and the relatively cheap land attracts industry, warehousing and retailing to peripheral sites. The increasing reliance on road transport for the movement of goods emphasises the benefits of a suburban location with its access to major roads such as motorways and ring roads. Recreation may require extensive sites, so golf courses and leisure complexes can be found in or on the edge of suburbs.

Which groups of people tend to live in suburbs and why?

The type of house ownership (tenure) that dominates in suburban areas is owner-occupation. People need to have a high enough income to obtain a loan to buy their house and so the occupational structure in the suburbs is dominated by the professional and managerial occupations. Many suburbs have very small proportions of their populations from ethnic minorities, as these groups may not have the economic security to afford the more expensive, lower-density housing. The situation is, however, dynamic: as migrants become established and their social and economic status improves, they too become an integral part of suburban communities.

There are large local authority housing estates on the edges of some towns and cities. The earlier ones date back to the 1930s and the more recent developments were built in association with the inner area slum clearance programmes of the 1960s and 1970s. Here, people with fewer economic resources can rent housing, although since the extensive sales of local authority housing in the 1980s, much of this has passed into owner-occupation.

How does urban growth create problems of urban sprawl and congestion?

The suburban expansion of the interwar years, along increasingly used roads and with infilling between the arterial routes, led to a growing concern about urban sprawl. The planning response was to try to constrain growth by enforcing green belts around towns and cities. In some locations, this has maintained a green area at the edge of the urban settlement, such as around London and Oxford and between Birmingham and Coventry, but in areas of high demand for housing, such as the southeast, urban sprawl has tended to leapfrog the green belt. Towns and villages just beyond the protected zone have undergone considerable pressure of development from people wanting to live at lower density in larger houses, yet continuing to have ready access to the urban centre for employment, retailing and entertainment, for example.

Developers prefer to build on greenfield sites at the edge of urban areas, as they are easier to build on. Population is increasing at a relatively slow rate, but with people living longer, more young people staying single for longer and marriages ending in divorce and thereby creating two households, the demand for houses continues to rise. There are, however, drawbacks: the loss of open space and productive countryside; increased use of cars for commuting and journeys to shops, for example; and longer journeys to work. There is, therefore, a link between urban sprawl and increased traffic congestion.

In other MEDCs, such as many countries in Europe, green wedges are used, allowing corridors of growth with protected areas in between. In the USA, very few attempts have been made to contain sprawl, as rates of car ownership are much higher and there is not the shortage of space experienced by many regions in Europe.

How does urban growth create problems of air pollution?

Much air pollution originates from the burning of fossil fuels. This is not a new phenomenon: London was subject to smoke and sulphurous fumes in medieval times from the large number of fires burning wood and coal. With the large-scale urban growth of the nineteenth century, dense fogs were generated in most urban areas. London's 'smog', a combination of smoke and fog, became infamous. Some 4000 people died in London during a particularly severe smog in 1952.

The reduction in coal burning for electricity generation and in factories and homes has resulted in a change in the types of air pollution commonly associated with urban areas. The air pollution tends to be less visible than the coal-based smogs, but nevertheless is a considerable risk to health. The combustion of diesel and petrol in vehicles produces emissions such as nitrogen oxides, sulphur dioxide, carbon monoxide and carbon dioxide, and microscopic particles. In cities where temperature inversions are common, smogs can be very persistent. The Los Angeles conurbation is particularly subject to photochemical smog caused by the reaction between oil-based emissions and sunlight, although conurbations such as London and Paris also suffer. The production of acid rain when sulphur dioxide, nitrogen oxides and water vapour combine has an immediate impact on the urban area in the accelerated weathering of building materials and damage to plants, and can affect regions beyond the built-up area.

How does urban growth create problems of waste disposal?

All human settlement produces material that is left over after various activities, such as manufacturing and domestic living. Over the past 200 years, the scale and, more recently, the type of waste produced by urban areas has changed. The rapid growth of towns such as Liverpool in the nineteenth century led to a concentration of polluted water, so that diseases such as cholera and typhoid frequently reached epidemic proportions. It is no coincidence that the first Medical Officer of Health was appointed in Liverpool by the mid-nineteenth century. Today, considerable efforts are made to deal with waste water from domestic sources and industry. Waste includes solids floating in the water and chemicals such as detergents and fat from food, the latter congealing to form large masses that can block drains and sewers. Water that has been used for cooling in an industrial process, if it is not allowed to cool before being returned to a river, can cause algal blooms and a loss of oxygen in the water that might result in eutrophication.

One of the greatest concerns facing many urban authorities is how to deal with the increasing volumes of waste from dustbins. A wide range of materials, for example plastics, metals, paper and organic matter, is thrown away each week, while larger items, such as fridges and televisions, add to the problem. The traditional method of disposal has been landfill at sites such as abandoned quarries in the fringe areas around urban settlements. However, this urban problem has been moved into rural areas, with the ever-increasing need to find more sites capable of swallowing vast quantities of waste. Attempts to incinerate waste create further issues of air pollution, especially in locations upwind of the plants.

How does urban growth create problems of water supply?

There is a basic need for clean water for domestic and industrial purposes. As urban areas have grown, so the scale of water management needed to supply hundreds of thousands of people has grown. Frequently, local sources such as rivers and groundwater are insufficient to supply urban water demand. This has led to urban authorities constructing reservoirs in rural regions to store water before transporting it to urban centres. Where groundwater is available, its use over many decades can result in the lowering of the water table, thereby causing environmental damage. The development of conurbations in semi-arid regions such as the southwest USA poses particular problems. For example, along the Colorado River and within California, dams, reservoirs and pipelines have been constructed to supply water to growing conurbations such as Phoenix and Los Angeles.

Urban land use and population patterns in cities

Key questions

Topic	Detail	Key questions
Urban land use and population patterns in cities	Patterns of urban land use in MEDCs and LEDCs, and urban morphology	• What are the main land uses within urban areas? • How is land use arranged in MEDC cities? • How is land use arranged in LEDC cities? • How do economic forces affect land use? • How do political forces affect land use? • What influences can physical factors have on land use patterns?
	Population distribution and density	• How and why is population distributed within urban settlements?
	Spatial segregation of groups within cities; income, life cycle, ethnicity and housing stock	• What is meant by spatial segregation? • How does income affect where people live? • How does life cycle affect where people live? • How does ethnicity affect where people live?

What are the main land uses within urban areas?

In terms of area, residential land use dominates. The other main built land uses are retailing, offices, manufacturing and services such as hospitals and educational establishments. Land is also taken up with transport infrastructure such as roads and railways. There can be significant areas of open space within towns and cities, for example parks and sports grounds.

How is land use arranged in MEDC cities?

Several models of urban land use have been developed, but as with all models they are simplifications of reality. They are the product of the time at which they were produced, reflecting the economic, social and cultural conditions of that time and earlier. However, they are helpful in highlighting key processes and patterns, some of which can be found in the urban areas of today.

Many MEDC cities are focused on a central business district. The land use around this can be arranged as a series of concentric rings or zones. Cities that grew quite rapidly from a central point tend to have each new phase of growth added on the outside of the previous ring.

An alternative arrangement has land use radiating from the city centre as a series of sectors; within each sector, one type of land use predominates. These sectors tend to develop along a linear feature such as a river or a man-made transport route, for example a main road or railway. There is, therefore, a strong directional element in this arrangement. Within an individual sector, the older land use tends to be towards the centre, with more recent developments of the same type of land use built towards the outskirts.

The multiple-nuclei model describes a situation where there is more than one centre, with each one being the focus of a specialised land use. Industrial, retailing and office areas are typical of the types of land use. This model assumes a high degree of personal mobility, as in North America.

How is land use arranged in LEDC cities?

There is much diversity among these cities due to their contrasting historical development, including the different experiences of colonialism and present-day economy and culture. However, some common features can be recognised. A commercial core area is quite common, as many of the larger cities were developed on the basis of trade; often, this is associated with port functions. If there was a strong colonial influence, walled forts and clearly defined areas of low-density housing for the administrators can be identified. The housing was often taken over by indigenous government officials after independence. Traditional commercial centres, such as bazaars, are common in areas of unplanned higher-density housing, usually located away from the higher-status areas. More recent industrial developments are generally found at the edge of the built-up area or close by the port, where squatter settlements develop to house those working in the factories.

How do economic forces affect land use?

Where there is a free market for land, whoever is prepared to offer or bid the most money will usually acquire the land. There is a general trend of decreasing land cost with increasing distance from the city centre. Some commercial land users are prepared to bid more for the centre than other potential users, as this central location gives the best access to customers. Traditionally, the city centre has been the most accessible location relative to the rest of the urban area. Transport routes converge on the centre with the built-up area spread more or less evenly around it. Once this location operates as the centre, it acquires other advantages, such as a certain type of building and prestige and specialist services. Residential land use cannot compete with commercial use for the central locations and so bids for land further away. Many cities have grown as housing has outbid agriculture for land on the outskirts of the urban area.

Within residential areas, there are clear differences according to cost of housing. In general, there is an increase in housing cost with increasing distance from the centre. This leads to the apparent paradox of cheaper houses in the inner areas being located on more expensive land and the reverse occurring in the outer suburbs with expensive houses on cheaper land. The solution to this paradox lies mostly with the density at which the housing is built. The inner areas are built at much higher densities than the suburbs, where most houses have some land around them in the form of gardens and driveways. There are exceptions, as with the high-cost locations towards the centre. Sometimes, these are long-standing patterns: for example, in London, parts of Kensington and Chelsea retain high-status housing as these neighbourhoods benefit from being close to the high-order functions of government and the City of London. Other inner areas have been gentrified — that is, an influx of higher social status groups has bought run-down properties and renovated them. There may be lower-value areas on the outskirts in the form of local authority estates.

How do political forces affect land use?

In some circumstances, land does not have a monetary value and cannot be bought or sold. Many eastern European countries followed a communist/centrally planned system for much of the second half of the twentieth century, under which land was owned by the state. Urban developments were not subject to market forces and other considerations could be seen in land use, such as the construction of large-scale public housing projects and areas of heavy industry.

Many MEDCs have a mixed economy of private and state influence. Under such a system, public housing can play a significant part in urban development. Most UK cities have extensive areas of local authority housing on their outskirts, or in areas of comprehensive redevelopment in the inner city. In the 1980s and 1990s, much publicly owned housing was sold and so the amount of housing available for rent from local authorities fell. Some other land uses are also publicly owned, such as hospitals, schools and parks.

The planning process represents a political force as well, as when a green belt is placed around a settlement.

What influences can physical factors have on land use patterns?

When an area is covered by buildings, roads and railways, it can be difficult to see the influence that the physical landscape can have. Ordnance Survey maps at 1:50 000 and 1:25 000 scales can reveal some of the physical factors.

If an urban area is located on the coast, it can only develop inland from and along the coastline. This will modify the arrangement of land use, so that if a series of concentric zones develops, the zones will exist as half-circles. The presence of a river may encourage sectors to develop along it. Floodplains tend to repel development, although with increased flood prevention in MEDCs, some land uses, such as industry, have been attracted to the cheaper land in such locations. In MEDCs, steep slopes also repel developments, although hills within urban areas often attract higher-status housing. There is increased competition for the sites that offer good views and a chance to be above the pollution of the rest of the city.

In many UK towns and cities, there is a marked west–east contrast in housing areas, locations to the west being more favoured and therefore more expensive than those in the east. The prevailing westerly winds blow the pollution generated by the urban area from west to east.

In LEDCs, the poor often have little choice but to occupy land that is marginal. Examples of such areas are tidal swamps, floodplains and steep slopes.

How and why is population distributed within urban settlements?

In most MEDC and LEDC cities, as distance from the centre increases, the population density decreases. There is, however, a significant difference between the two groups of cities.

In MEDCs, urban population density profiles have been reducing their gradient through time as people have increased levels of personal mobility and have decentralised to the suburbs and beyond. Inner-city redevelopments have often involved the demolition of terraced housing, in which people lived at high densities, and its replacement by new housing that allows lower population densities. Many CBDs have expanded into former residential areas, so lowering population density at the centre. The highest densities are in the inner suburbs, where terraced housing, tenements and flats tend to be the main housing styles. The outer, low-density suburbs are dominated by detached and semi-detached houses, most of which have some land front and back.

In LEDCs, population densities have traditionally been at steeper gradients, with higher central densities. As mobility within the city is restricted due to less efficient transport infrastructure and lower incomes, people need to live closer to goods, services and jobs. Higher-income groups often live close to the centre of the urban area in apartment blocks. At some peripheral locations, the presence of squatter settlements can give very high population densities. Through time, the city grows outwards, but usually at a similar gradient as personal mobility increases at a relatively slow rate.

What is meant by spatial segregation?

Where people live within an urban area is sometimes seen in terms of different social groups. There are a number of ways of placing people into social groups, such as income or ethnicity, and when these are put together, the pattern that results is in the form of a residential mosaic. Different social groups tend to be segregated from each other and live in different areas of the settlement.

How does income affect where people live?

Individual decisions that together make up the pattern of spatial segregation are strongly influenced by the level of income that people have. In the UK, nearly two-thirds of all households are owner-occupiers — that is, they either own outright or are buying their home. To achieve this, most people need to borrow money in the form of a mortgage and their salary largely determines how much a bank or building society is prepared to lend. Some occupations, such as professional and managerial jobs, are considered a lower risk by lenders.

Higher income also allows people to spend more on mobility, such as private car ownership, and to afford the higher costs of commuting from the suburbs.

How does life cycle affect where people live?

Many people move through a life cycle that involves changing the type of accommodation they need and therefore changing where they live within an urban settlement. Such moves are often associated with changes in income levels with age and/or with changes in household size, for example at marriage or on the birth of children. When young people leave home to set up their first independent household, they usually have limited income and do not need much space, so they often live in a relatively cheap, central flat. A couple with children might try to buy a larger house with more space and a garden in the suburbs, while in retirement the demand for space is much reduced and some people move to smaller accommodation or migrate away from the urban centre.

Stage in life cycle	Housing needs/aspirations
(1) Pre-child stage	Relatively cheap, central-city apartment
(2) Child bearing	Renting of single family dwelling close to apartment zone
(3) Child rearing	Ownership of relatively new suburban home
(4) Child launching	Same area as stage 3 or perhaps move to a higher-status area
(5) Post-child	Marked by residential stability
(6) Later life	Institution/apartment/live with children

Housing needs at different stages of the life cycle

Not everyone follows the cycle, and the high rates of divorce in many MEDCs result in a more complex set of housing needs. There are also locational differences throughout the stages, depending on whether the household is high- or low-income.

How does ethnicity affect where people live?

Within urban settlements, many ethnic groups are highly segregated. Generally, the more different the group is from the majority population, the greater the degree of segregation. Where the segregation is very pronounced, ghettos can develop.

Segregation develops due to a set of factors, some of which are positive while others are negative. Positive reasons include peoples' desire to share their home language, to share in their own places of worship and schools, and to have access to specialist shops, such as for particular foods. Negative reasons include defence and security against prejudice from the host society, a low economic status allowing little choice in the housing market, except for low-cost areas such as inner-city terraced housing, and discrimination in the housing market.

Over time, some ethnic groups are assimilated with the host society and so the individual households tend to disperse away from the original cluster.

Questions & Answers

This section contains six examination questions, two for each of the three topic areas outlined in the Content Guidance section: population, rural settlement and urban settlement. These questions have a short-answer format and are based around stimulus material such as diagrams, tables and maps. However, some parts give an opportunity to write at greater length, about a side of A4, on a place you have studied. Allow around 20 minutes for each question. The length of your answers should be proportional to the marks available for each part of the question.

Examiner's comments

All candidate responses are followed by examiner's comments. These are preceded by the icon *e* and indicate where credit is due. In the weaker answers, they also point out areas for improvement, specific problems and common errors such as lack of clarity, weak or non-existent development, irrelevance, misinterpretation of the question and mistaken meanings of terms.

Examiners use two marking strategies when assessing short-answer questions. It is important that you can identify how each question will be marked. Questions that require a simple statement, such as a definition or description and brief explanation, are usually point-marked. A typical point-marked question might ask you to state and explain one factor that influences the decline in mortality in LEDCs. An examiner would award a maximum of 1 or 2 marks for an accurate reference to improved water quality, and a maximum of 2 or 3 marks for an accurate and full description of how this would help reduce death rates, perhaps infant mortality in particular.

A different method of assessment is used for extended-answer questions. These questions are Levels-marked and are worth between 4 and 10 marks. The examiner's mark scheme usually defines two or three Levels, each Level containing a description of the qualities of a typical answer.

A Level 3 answer demonstrates good knowledge and understanding of a specific scheme and a sound appreciation of its effectiveness. A Level 2 answer, for instance on the relationship between population size and the number of goods and services available in a settlement, might be accurate, focused and show a clear understanding of how population numbers (threshold) influence how many shops and services are able to locate in a settlement. In comparison, a Level 1 response might contain some understanding, but will generally lack focus and will make an unconvincing connection between population and functions.

In the longest answers, worth 10 marks, candidates who write accurate, general answers, but with little if any place-specific detail, gain a maximum of 4 marks. It is also important that knowledge and understanding of a particular location are used to answer the specific question set: these questions are not opportunities for a candidate to write all they can remember about the place.

Population (I)

Figure 1 shows crude birth rates and crude death rates for a sample of Asian and European countries in 1999.

Figure 2 is a dispersion diagram showing the crude birth rates for the same sample of Asian and European countries in 1999.

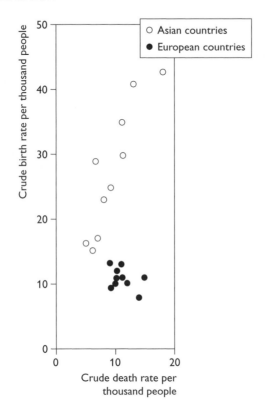

Figure 1

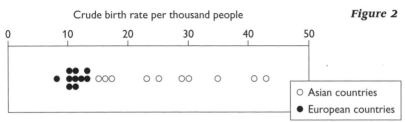

Figure 2

(a) What is meant by the term 'natural increase'? (2 marks)
(b) (i) State the overall trend in the relationship between crude birth rate and crude death rate among the Asian countries. (2 marks)

AS Geography

question 1

(ii) **Using evidence from Figures 1 and 2, describe the differences in crude birth rates and crude death rates between Asian and European countries.** (6 marks)

(iii) **Suggest reasons for the differences in crude birth rates.** (6 marks)

(c) **State and explain the possible influence of two factors causing variations in the death rates among Asian countries.** (3+3 marks)

Figure 3 shows population pyramids for a typical **MEDC** (pyramid A) and a typical **LEDC** (pyramid B).

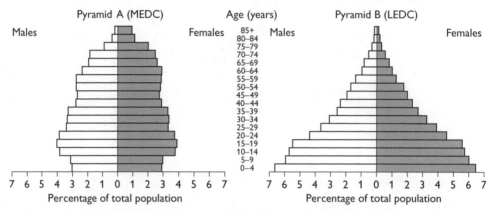

Figure 3

(d) **With reference to Figure 3, identify and explain the differences between the two pyramids for the 15–64 age group.** (5 marks)

(e) **With reference to either a named MEDC or a named LEDC that you have studied, suggest two possible consequences arising from its population structure for the 0–14 age groups.** (4+4 marks)

Total: 35 marks

■ ■ ■

Answer to question 1: candidate A

(a) Natural increase is the difference between the crude birth rate and the crude death rate.

   ✏ This is a precise and accurate answer, worth 2 marks.

(b) (i) From Figure 1, there is a positive trend between the rates. As crude death rate increases, so does crude birth rate.

   ✏ This is another precise and accurate answer, clearly using Figure 1, for 2 marks.

   (ii) With regard to birth rates, the difference between the two areas (Asia and Europe) is clear. Asian countries have greater birth rates (ranging from 14 to 43 per thousand people) whereas European countries have lower birth rates (ranging from 8 to 13 per thousand). Asian countries have a greater range in death rates (from 5 to 18 per thousand), but Europe has a narrower range from 9 to 15 per thousand.

> 🖉 This is a detailed and accurate description, in which the candidate uses the resource to good effect. Both regions are mentioned and figures quoted. The question asks for the differences between Asian and European rates to be highlighted, which this candidate does clearly. The answer is worth the full 6 marks.

(iii) In Asian countries, which are often LEDCs, children are seen as economic assets as they can work from a young age. They can help on the farm or in a business. In Europe, children are more of an economic drain as they stay at school until they are 16. Also, contraceptive devices are less readily available in Asian countries than in Europe. In addition, more babies are born to compensate for the high infant death rate in Asian countries.

> 🖉 Three reasons are given here, all concerning birth rates. They are appropriate reasons and stated well enough to be given the full 6 marks. The contrasts between Asian and European countries are clear.

(c) Some Asian countries have less pollution than others and so death rates are lower. Some Asian countries will have more disease than others so that more people will die and so death rates will be higher. The sewage systems in some countries, such as Taiwan, are better and so less disease will be spread.

> 🖉 The first answer correctly states a possible factor, for 1 mark, but there is no explanation of this. The second factor is also a sensible suggestion, which is developed, for 3 marks.

(d) The population pyramid for the LEDC has a very wide base, showing the high birth rate. However, in the older sections of the population the LEDC pyramid decreases in size, whereas the MEDC one remains fairly constant. In the 20–24 age group and upward to 64 there are a greater number of people in the MEDC.

> 🖉 This answer starts off describing part of the pyramid not required by the question. From then on the answer describes the differences quite well, but no explanation is offered, so only 3 marks can be awarded. It is important to read the question carefully so that irrelevancies are avoided. If there is more than one command word (in this case identify and explain), make sure you answer both.

(e) In Britain (an MEDC) the number of people in the 0–14 age bracket is very low compared with the high numbers of elderly. This means that in a few years, when the under-15s are working, there will be a high dependency ratio and taxes will have to be increased to support this greying population.

If there are fewer babies being born, this could cause a decline in the population of a country. A small population would make the country very vulnerable and easy to attack if a war broke out. Countries are often afraid of a declining population.

> 🖉 The first answer is good, as it suggests a sensible consequence for an MEDC of a small number of under-15s. It gives a sound explanation of the economic impact. The second answer is interesting, as it is not one of the main consequences that

49

students tend to write about. Examiners are told to look out for such answers and to be prepared to give credit where this is due. However, you often do better to focus on the more straightforward answers. This student does not develop the point fully and the answer probably needs an example to be really convincing. 4 marks are awarded for the first part of the answer, 2 marks for the second.

e **Overall, candidate A scores 29 marks out of 35. This is a good A-grade answer.**

■ ■ ■

Answer to question 1: candidate B

(a) This is the number of people who are born, minus the number of people who die per year in a place. No migration is allowed.

e Although expressed rather simply, this answer nevertheless has the essence of the definition and so is worth 2 marks.

(b) (i) Overall, as the birth rate increases, so does the death rate. However, the birth rate increases twice as much as the death rate.

e This is a good answer. It is concise and accurate with a comment about the nature of the relationship, and is clearly worth 2 marks.

(ii) In Asian countries, the crude birth rate is quite high in relation to the figures for the European countries. In Europe, the birth rate is relatively low. In Asian countries, the crude death rate is also higher than in European countries but not by as much. There are some Asian countries with a death rate lower than in Europe.

e In terms of the differences, this candidate has given a very encouraging response. Both birth and death rates are considered and the candidate has picked up on the interesting aspect of lower death rates in some Asian countries than in some European ones. The only deficient aspect of this answer is that no evidence is quoted from the figures and the question is quite clear that such evidence, i.e. data, is required. This answer is therefore worth 5 marks.

(iii) Asian countries tend to be less economically developed as they have more agriculture; this tends to be subsistence farming, such as rice paddies. They have high birth rates because there is less education about contraception, and therefore many don't use it. They also have more children because there is a high infant mortality rate. They do not have pensions and so children need to look after their parents when they are old.

e The points this candidate makes are valid, although the first sentence needs to make the link between subsistence farming and children being able to contribute to the family farm. In the rest of the answer, the link between higher birth rates and factors such as family planning and infant mortality is quite well made. There is, however,

OCR (A) Unit 2

no explicit mention of why these Asian rates are higher than the European ones, so this is not as complete an answer as it might be. It is worth 4 marks.

(c) One country may have progressed further on the demographic transition model to a stage where deaths are low. Other countries may be at an earlier stage where deaths are high.

Health services will be generally better in more wealthy countries such as Japan. Countries that are not as wealthy, for example India, will not have such good health care.

e The first answer is not a factor and in some ways is repeating the question but using different words. No credit can be given here. Factor 2 is very good as it answers the question directly, using two well-chosen examples to support the point. 3 marks are awarded for this answer.

(d) In the LEDC, the 15–64 age groups tend to decrease significantly as you get to an older age. This could be down to two possibilities: lots of deaths and out-migration. There are higher death rates in LEDCs compared with MEDCs and many young people migrate to MEDCs for jobs. The MEDC has very gently tapering sides in the 15–64 age group.

e There is both description and explanation here of the LEDC pyramid, but only description of the MEDC one. The descriptions are fine, as they deal with the actual shape. The points made about the LEDC are plausible, although the impact of out-migration is not likely to be that significant for an entire country unless it is a very small one. The candidate should have made some attempt at explanation of the MEDC pyramid. Overall, this answer is worth 3 marks.

(e) In India, the population is over one billion as a result of high birth rates. Much of the population is under 15 and so enormous pressure is put on the health care facilities. When these people are older they will have families of their own and even more pressure will be put on health care. They will also need houses and jobs.

LEDCs do not have pension schemes and so people will have to work longer to support their families. They will probably have to find work in the informal sector to support themselves.

e The first answer is good. The point made about health care is correct and clearly expressed. The last sentence is not needed, as it introduces two more consequences. One of these should have been developed as the second consequence because the point about pension schemes is irrelevant. The absence of pension schemes is a consequence not of large numbers of under-15s but of the relative poverty of an LEDC. People do not have the disposable income to save for the future, unlike in MEDCs. No credit can be given for consequence 2, but 4 marks are awarded for the first answer.

e **Overall, candidate B scores 23 marks out of 35. This is a B-grade answer.**

AS Geography

Population (II)

Figure 1 shows population change in the European Union in 2000.

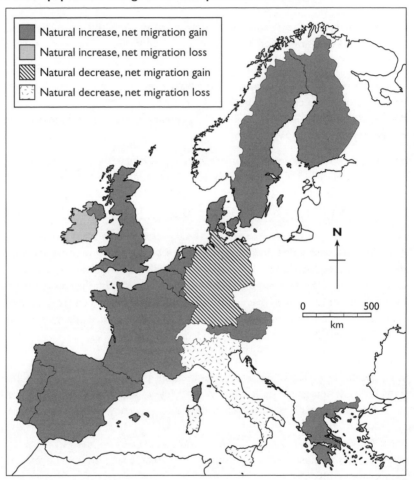

Figure 1

(a) Describe the pattern of population change in the European Union. (4 marks)
(b) State and explain *two* possible reasons for international migration. (3+3 marks)
(c) Describe *two* ways in which a migration gain into a region might alter its age–sex structure. (4+4 marks)
(d) State and explain how intervening obstacles might interrupt movements of migrants at an international scale. (6 marks)

Total: 24 marks

Answer to question 2: candidate A

(a) The majority of the European Union has undergone a natural increase with a gain in migration. Ireland's population has increased naturally with a loss in net migration. Italy and Germany have had a natural decrease in population but Italy has had a net migration loss whereas Germany has had a gain.

> 🖉 The basic description is sound and the candidate has also picked up some detail with the naming of countries that deviate from the general pattern. This is worth all 4 marks.

(b) The migrant country may not be politically stable and so inhabitants may be forced to leave in order to seek better standards of living.

People may be attracted by the opportunities that lie in a different country, for example job and housing opportunities. These may be easier to obtain and maintain in another country.

> 🖉 The candidate's weak expression does not help to convey the answer. Reason 1 introduces an appropriate factor — political stability — but the rest of the answer does not develop this point, so this is awarded 1 mark out of 3. Reason 2 is again a suitable factor and the candidate does add some comment about employment and housing, but not in a convincing way, so 2 out of 3 marks are given here, making a total of 3 out of 6 marks.

(c) There may be a large number of young people who have just left university looking for cheap housing. They would be attracted by small, cheap houses in areas where possibly retired people live, therefore reducing the number of older people relative to the number of younger people.

There may be a new factory that requires a lot of skilled workers. They would want to live near their work so they may move into the area inhabited by families. This would increase the number of people in an area where there were lots of males and females.

> 🖉 Neither answer goes beyond Level 1, although there are the beginnings of two sensible points. The first answer, migration of young people into a region for cheaper housing, is basically correct but is not developed and the link with retired people is suspect. The second answer also identifies a correct way, migration for employment, but again does not go on to describe convincingly its effect on age–sex structure. Ideally, this answer would include an example of such a population change, as this would confirm the candidate's understanding. This answer is worth 2 + 2 marks, giving a total of 4 out of 8.

(d) People may need to go to a country that is far away and hard to get to. They may need to cross a large stretch of water and if they don't have the means, then it may prevent them from reaching their final destination. Sometimes there are no ships that go across a sea, or it is too far to travel.

question

> *e* This is an appropriate point about a physical barrier, a sea or ocean, but it is not well made. There is some mention of potential migrants not having the means, but it is not clear exactly what the candidate is stating. In this Levels-marked answer it is best to give at least two appropriate points that are explained in full. This response leaves the examiner with an incomplete answer, so it receives a Level 1 mark of 2 out of a possible 6 marks.

> *e* **Overall, candidate A scores 13 marks out of 24, which is a grade-C answer.**

■ ■ ■

Answer to question 2: candidate B

(a) There has been natural increase and net migration gain in west Europe, northeast Europe and small amounts of south Europe. There has been natural increase and migration loss from one country and natural decrease and migration gain in another. The pattern is quite mixed.

> *e* The command to describe a pattern is common in this paper, but candidate B does not describe the overall picture. The answer simply lists areas where various demographic changes are found. It would be better to give a statement that summarises the common situation (in this case, that most of the EU has undergone natural increase with migration gain), highlight a couple of areas where that was true and then point out some of the anomalies, such as Germany and Italy. There are some correct elements in this answer and the candidate has used the resource for 3 marks out of 4.

(b) Push factors take people away from their homes, as they no longer want to stay there. If there was a war in their home, then people would try to leave.

Pull factors attract people to a different area in a voluntary way. There may be poor weather in their own country and better weather in another one abroad. There might also be better standards of living.

> *e* In both cases, the candidate states a valid reason but does not explain in a convincing way how it links with international migration. The phrase 'better standards of living' is vague and should be avoided. 1 mark each for the two correct reasons is awarded, giving 2 out of 6 marks.

(c) It is generally the young people who migrate as there are better prospects for their families in other countries. As they are young, they tend to have children, so the average age of the country they are moving to will change.

It is mainly young men who migrate to find wives and better jobs. The country they have left will have fewer males and so its age–sex structure will alter.

> *e* The focus of the question is on migration altering age–sex structure. In the first answer, although the basic migration type is sound, no clear outline of a change to age–sex structure is given. In the second answer, the candidate focuses on the origin

of the migrants, but the question asks for the impact on the area showing a net gain in migrants. So, in total, only 1 mark can be awarded.

(d) The borders of countries are often very hard to cross permanently. The migrant requires a passport or a visa and probably a good job to go to. These can be very hard to get in some countries, for example for Mexicans migrating to the USA. The social opinions of some people might mean certain other people won't move there as there would be racial attitudes towards them.

> ℯ The first part of this answer is good, as it mentions an appropriate intervening obstacle and explains how this interrupts international migration. The use of the example, although not required by the question, confirms the candidate's understanding of the point. The second part of the answer does not deal with an intervening obstacle, as racial discrimination is present in the destination and along the route of the potential migrants. As this is Levels-marked, and the first part of the answer was tackled well, 4 out of 6 marks are awarded.

> ℯ **The candidate scores 10 marks out of a maximum 24 marks. This is an E-grade answer.**

Question 3

Rural settlement (1)

Study Figure 1 and Table 1.

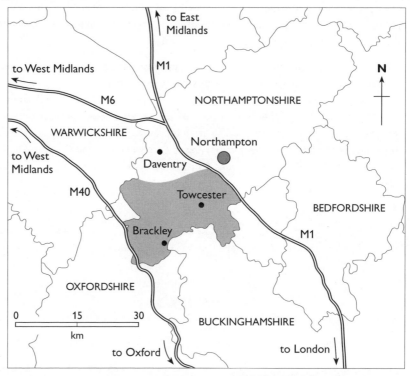

Figure 1 Part of the southeast Midlands

	Village population					
	Small (37 settlements) 0–299 population		Medium (33 settlements) 300–999 population		Large (19 settlements) 1000–3000 population	
Shop/service	Number	%	Number	%	Number	%
General store	2	5	14	42	17	89
Post office	5	14	25	76	18	95
Public house	12	32	29	88	19	100
GP surgery	1	3	3	9	12	63
Primary school	1	3	23	70	19	100

Table 1 Population size and shop and service provision in south Northamptonshire

OCR (A) Unit 2

(a) What is meant by the 'range' of a good or service? (2 marks)
(b) From the data in Table 1, identify *one* low-order function and *one* high-order function. (2 marks)
(c) Using the evidence in Table 1:
   (i) Describe the relationship between village population size and the provision of shops and services. (4 marks)
   (ii) Explain the relationship you have described. (4 marks)

Over the past 30 years, the number and variety of shops and services in rural areas such as the southeast Midlands has declined.

(d) Describe and explain *two* possible reasons that might have caused this decline. (5+5 marks)
(e) For a named rural region in an MEDC you have studied, describe and explain how changes in its service provision have affected some rural dwellers more than others. (10 marks)

Total: 32 marks

■ ■ ■

Answer to question 3: candidate A

(a) The range of a good or service is the distance the customer is prepared to travel to obtain it, for example a pint of milk has a small range, whereas an electrical item, such as a fridge, has a large range.

   e This is an acceptable definition, helped by the correct examples, and receives 2 marks. Given the time constraints of an exam, the opening phrase in the answer could have been omitted.

(b) Low-order function — general store
   High-order function — GP surgery

   e Two correct examples are given, so 2 marks are awarded. It is important that the examples come from Table 1, as the question asks, 'From the data in Table 1…'

(c) (i) As the village population increases, the provision and variety of shops and services also increases. Only 3% of the small settlements in Table 1 have a primary school whereas 100% of the large settlements have a primary school.

   e The clear and accurate description of the relationship in the first sentence is a good start. The question asks the candidate to use evidence from the resource and this answer does that in the second sentence. This is worth the full 4 marks.

   (ii) The provision of shops and services increases with population because when the population is greater it reaches the threshold value for more services.
   There is often not enough demand in small settlements for services such as a GP surgery or a shop such as a chemist to be economically viable.

AS Geography

question 3

> 🖉 This an encouraging answer: it deals directly with the relationship described in (c)(i), giving the correct explanation by using the appropriate term, 'threshold'. The candidate then sets this general statement in context by using the GP surgery as an example. For this precise response, full marks are awarded.

(d) One reason could be rural depopulation. People might be moving out of the rural areas to towns for employment and so the local thresholds have decreased. People might also be moving out because they cannot afford to buy a house, since if there are many second homes then house prices might be high. The people who buy second homes often have high incomes and can out-compete the locals.

Another reason for decline in shops and services could be the increased mobility of the rural population. With more people owning cars, they are able to travel to the supermarket in the local town to do their shopping rather than use the small village stores. This in turn means there is less demand for the local shops and so they have to close down.

> 🖉 When a question asks for more than one reason, it is better to give two quite different reasons. This candidate does this by describing rural depopulation and increased personal mobility. Both factors are then fully explained in the context of the decline of shops and services, showing that the candidate has read the question carefully and is giving a thoughtful response. Both answers deserve full marks.

(e) When the slate mines were closed in north Wales it brought much unemployment to the rural area. Because people were out of work, they could not afford to live there any more and so they left. They went to towns on the coast, such as Llandudno, where they could find work. Some left north Wales to go to large settlements such as Liverpool and Manchester and some even went overseas. Therefore, many shops and services were closed, as they did not have enough customers.

Some of the people who stayed living in the area were badly affected: if they were elderly and did not have a car, they had to shop in the few shops that stayed open. People with cars are able to travel to the local towns such as Llandudno to buy their goods more cheaply in the supermarket. The local bus services are very poor, as their threshold cannot be reached. This means that the elderly people who have always lived there are badly affected. There is a downward spiral as further depopulation of the area leads to a further decline in services.

> 🖉 The extended answer requires the candidate to make use of a place-specific example. The question is Levels-marked. Candidate A chooses a sensible place on which to base the answer, as it is clearly a rural region. The first part of the answer describes and gives reasons for depopulation, which is not required by this particular question. The second half of the answer does refer directly to the question, focusing on the impact on the elderly and drawing a contrast with those who are more mobile via their cars. The essence of the question is the different effects of changing service provision on different groups within the rural area. The

OCR (A) Unit 2

candidate has not manipulated sufficiently well the knowledge and understanding they have of their chosen rural region to answer this particular question. It is important that answers to this style of question are not taken as opportunities for candidates to write all they can remember about the case study. This answer receives a Level 2 mark of 6 out of 10.

☑ Candidate A scores 28 out of 32 marks — a high-quality A-grade answer.

■ ■ ■

Answer to question 3: candidate B

(a) The area which a good or service gets sold to. It is the marketing area of that good or service.

☑ This answer is inaccurate, so no marks are awarded. The candidate does not appear to understand the term 'range', which refers to the *distance* people are prepared to travel to obtain a good or service.

(b) Low-order function — post office
High-order function — GP surgery

☑ Two correct examples are given, so 2 marks are awarded.

(c) (i) The larger the settlement, the greater the number of both low- and high-order functions. As the size of the settlement increases, the percentage of shops and services also increases. For example, small settlements only have one GP surgery, whereas medium settlements have three and large settlements have 12.

☑ Although not well expressed, this answer contains a correct description of the relationship. The first sentence has the essence of the pattern while the rest of the answer gives evidence for the relationship from the resource, as requested by the question. This answer is worth 4 marks.

(ii) The larger the settlement, the greater is its population size, and therefore the more of each service the settlement will need. The larger the settlement, the fewer settlements there are and the more money the council gives to them. This means that they need more goods and services.

☑ The first sentence contains hints at the basic explanatory point — that of threshold — with the link between population and number of services. It is not well expressed and is not developed, but can be awarded 2 marks. The remainder of the answer is very confused and adds nothing.

(d) Migration to urban areas. The more people move from these rural areas to urban areas, the more the need for services decreases. Young people tend to move, so there is less need for primary schools as there are fewer families with children. There are also fewer customers for the shops and so they go out of business.

Local councils have been focusing their attentions and money on the urban areas and increasing the appeal of these areas to the young. They have been ignoring the rural villages, which have been left to decline, with no young residents to start up new businesses.

> For the first reason, a correct point is made about possible rural depopulation via out-migration reducing the demand for shops and services. The focus on the loss of younger people is well linked with the drop in demand for, and therefore provision of, primary schools. This is worth all 5 marks. The second reason indicates some muddled understanding by the candidate. An answer can focus on settlements in a rural area, such as larger villages, as part of a key settlement policy, but the reference to urban areas is inaccurate and mention of younger people is not helpful. In those areas where a key settlement policy has been used, the villages not chosen as key settlements often do experience a decline in shops and services. No credit can be given here.

(e) Due to counterurbanisation, there has recently been an increase in the population of parts of Suffolk. This has caused an increase in the number of houses in the area. The new residents think the services need changing but the older residents think that the services were better in the past when there were more shops. There is now great tension between these residents, which has caused conflict.

The new residents in some villages are mainly young couples with children and the old residents are mainly elderly people. So the new services which have been supplied are aimed at the younger generation, for example primary schools and youth clubs, which the elderly people do not need. The elderly are not as mobile and cannot get to the local towns for their shopping as easily as the younger people can.

> An appropriate rural region is used, but the only reference to an actual place is the county name. Extended answers are expected to show evidence of detailed knowledge of a particular area, such as village names and perhaps some statistics highlighting the changes. The first paragraph in this answer is muddled and peripheral to the question and gains the candidate no credit. The second paragraph has some useful material relating to the question but, overall, only a Level 1 mark of 4 out of 10 can be awarded.

> **Candidate B scores 17 out of 32 marks for this question. This is just of D-grade standard.**

Rural settlement (II)

Study Figures 1 and 2.

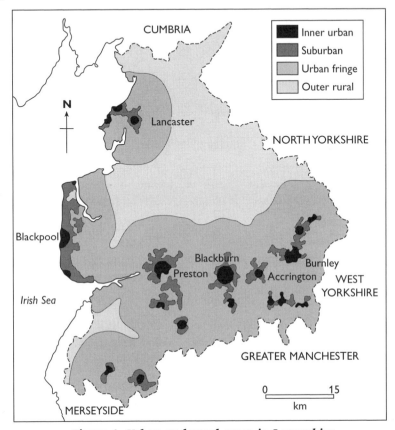

Figure 1 Urban and rural areas in Lancashire

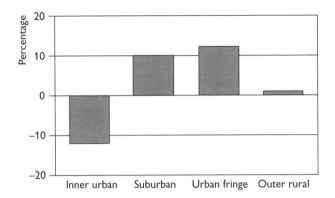

Figure 2 Population change in Lancashire, 1971–91

AS Geography

question 4

(a) Describe the pattern of population change within Lancashire. (4 marks)
(b) What is meant by the term 'counterurbanisation'? (2 marks)
(c) How might the process of counterurbanisation help to explain some of the population changes in the urban fringe? (6 marks)
(d) State and explain **two** reasons why counterurbanisation appears to have had little influence in the outer rural areas. (3+3 marks)
(e) State and explain **two** differences in the likely age structure of *outer rural* and *urban fringe* areas. (3+3 marks)

Total: 24 marks

■ ■ ■

Answer to question 4: candidate A

(a) The inner areas declined but the rest of Lancashire increased, although the outer areas increased very little. The inner areas declined by about 12%. The two areas that gained most were urban fringe (12%) and suburban (10%). The outer rural hardly changed.

> *e* This is a detailed and accurate description. The overall pattern is described, followed by the detail of the different areas. The candidate uses the resource very well — quoting figures to support the answer. 4 marks are awarded.

(b) This is when people move out of urban areas to the rural areas. The rural areas increase in population, as this is where people now want to live.

> *e* Although this is not a particularly precise definition, there is enough evidence of a sound understanding of the term to gain 2 marks.

(c) When people leave urban areas, they go to villages and small towns. They sometimes move to live in a larger house on cheaper land away from the pollution. As they still work in the town, they need to commute.

> *e* This has the basis of a good answer, as it focuses on migration out of urban areas to the fringe and there is mention of some of the reasons why people move. However, this is typical of many candidate answers that imply a link between factors and patterns rather than making it explicit: for example, the link between the need to commute to the city and living in the fringe, which is not too far to travel. This is a Level 1 answer, worth 4 marks.

(d) The outer rural areas do not have good roads to allow people to travel to the towns as quickly as they need. Also, they do not have railways to help them commute to work.

Most of the outer rural areas do not have shops or services such as schools and doctors. People coming from towns are used to these and will not want to move to an area that doesn't have them.

OCR (A) Unit 2

e Both these reasons are valid. The development of both also relates to the out-migration of people from the urban areas, which is the core of counterurbanisation. Full marks are awarded.

(e) In outer rural areas, there is out-migration of young people because there are not so many jobs in farming today, so there are fewer 20- and 30-year-olds. In the urban fringe, there are more people of these ages because they can find work in the town.
In outer rural areas, there are more elderly people because they like to retire to the peace of the countryside. The urban fringe is more busy and this is where families live.

e Both answers are valid, with the first one in being well explained, for full marks. The second answer concerning the urban fringe is less well expressed — 'more busy' is a poor phrase. A comment about the provision of shops, schools and health care facilities would be more convincing, so 2 out of 3 marks are awarded for the second answer.

e **Candidate A scores 21 out of 24 marks, which is a very good A-grade answer.**

■ ■ ■

Answer to question 4: candidate B

(a) The different parts of Lancashire had different types of population change. The inner urban areas went down but all the others went up, although the outer rural did not go up by much.

e This answer has the basic idea of a variable pattern and mentions a couple of the different areas by name with the correct type of population change. There is, however, a lack of detail, such as citing the figures, so the answer receives 2 out of 4 marks.

(b) This is when the urban areas stop growing as no more people are moving into them.

e This is an inaccurate answer. The shift in population from the larger urban settlements to rural areas must be emphasised. No marks can be awarded.

(c) In the fringe, the population is increasing because people are moving to these areas to get away from the town. People want to live in pleasant surroundings and have a better standard of living. Young families in their life cycle move to the fringe to bring up families in more space.

e The candidate has the idea of out-migration from urban areas by certain groups in particular. This is explained in terms of residential preference as regards environment. There is no mention of the fringe being within commuter range of the town, however, or of the fringe having the highest population change among the four types of area. 4 out of 6 marks are awarded.

AS Geography

question

(d) The outer rural areas have only increased a little as they are farming areas and there are fewer jobs in farming now. People don't want the hassle of working out of doors all the time in bad weather. Farming is now a struggling business.

There aren't as many settlements in this area and so there aren't as many places to live in. Some of the villages are very small and there may not be houses for people, so they have to move out.

> *e* The first reason is not concerned with counterurbanisation; rather it deals with issues to do with the agricultural sector, so no marks are awarded. The second reason has some merit regarding the lack of housing in outer rural areas, but it is mentioned in the context of out-migration from the rural region. If the candidate had put this point over as restricting the availability of housing for potential incomers from the urban areas, more credit could have been given. As it stands, the second reason receives 2 out of 3 marks.

(e) In outer areas, most people are likely to be elderly as the young people have moved to the towns for jobs and entertainment. The elderly will be retired and therefore won't mind the peace and quiet.

In the urban fringe areas, there are more families with young children than in the outer rural areas. This is because they have moved to the fringe where there are schools and shops. In the rural areas, many schools and shops have closed down as there are not enough people.

> *e* The first answer gives a valid point about age structure, although the contrast with the urban fringe is implied rather than stated clearly. The comments about the out-migration of young people and the presence of retired inhabitants are also valid, so the first answer receives 2 marks. The second answer is accurate and contains sufficient explanatory material to be awarded full marks.

> *e* **Overall, candidate B scores 13 marks out of a maximum 24. This is a D-grade answer. This candidate would have scored significantly higher with a better knowledge of the process of counterurbanisation and a more careful reading of the question.**

Question 5

Urban settlement (I)

Figure 1 shows Dar es Salaam, the capital of Tanzania, east Africa, and its largest urban settlement. Its population is growing rapidly at about 10% per year. The majority of the inhabitants, some 60%, live in squatter settlements.

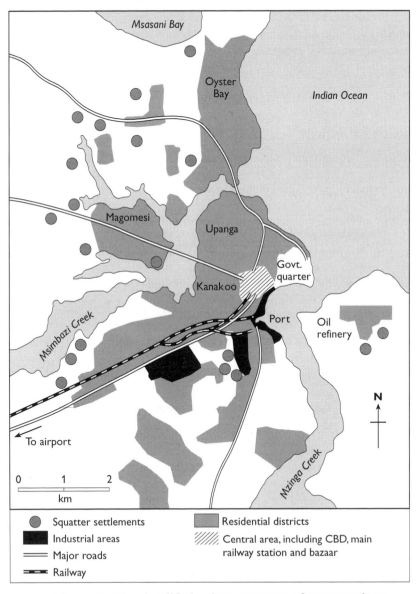

Figure 1 The simplified urban structure of Dar es Salaam

AS Geography

question 5

(a) **Describe the distribution of squatter settlements in Dar es Salaam.** (4 marks)
(b) **State and explain *three* possible reasons for the siting of squatter settlements.** (3+3+3 marks)
(c) **Give *two* possible reasons for the growth of squatter settlements in most large urban areas in LEDCs.** (3+3 marks)
(d) **Using evidence from Figure 1, suggest one possible reason for the development of the residential sector that extends in a southwesterly direction from the central area towards the airport.** (3 marks)
(e) **For a named city you have studied, describe and explain how recent changes have caused environmental problems.** (10 marks)

Total: 32 marks

■ ■ ■

Answer to question 5: candidate A

(a) The squatter settlements tend to be situated near the main residential districts. Some settlements are near industrial areas, a few are near major roads which lead to residential or industrial areas, and the occasional settlement is located slightly further out of town, by water.

e The candidate keeps to description, when a common mistake is to include explanation. The description is rather limited, although the candidate has made some sensible observations. It is best to try to open with a statement that summarises the overall pattern, then offer some detail, for example a place name or other reference from the map, and then perhaps point out any anomalies. In this case, the answer is awarded 2 marks.

(b) Near industrial areas, the squatters can easily get to jobs in factories, which use a lot of labour. They cannot afford a long journey.

Next to residential areas, many jobs are available. These are fairly low-paid, but even children can get jobs. Also, schools are available for children.

Further out of town, near water, some residents can grow their own crops and others can walk into the residential or industrial areas for jobs. These settlements are less polluted and so are better to live in.

e When more than one reason is requested it is important that clearly different reasons are given. It is also important to state clearly and then suggest how each particular factor operates. The first reason stated, access to employment, is worth 1 mark. Its explanation is valid and worth the 2 remaining marks. The second reason duplicates the first in referring to jobs, but 1 mark can be awarded for the mention of access to schools for children from the squatter settlements. The final reason is a confused response, containing invalid points, and so receives no marks. 4 marks out of 9 are awarded.

(c) Poor climatic conditions in rural areas make growing crops difficult and so rural dwellers move to urban areas in search of jobs.

When the migrants arrive, they have very little money and cannot afford the prices of houses, so they settle in squatter settlements.

e The first reason is an example of a candidate not reading the question with sufficient care. The question asks why squatter settlements have grown, not why people move to urban areas, so no credit can be given here. The second reason is appropriate as it relates directly to the issue of affording somewhere to live and how many urban dwellers have little choice but to live in squatter settlements. This is worth 3 marks.

(d) Settlement has grown along the major road between the CBD and the airport. At both ends of the road there are job opportunities, so people want to live along the road so they can get to these easily. Although not many people will have cars, there are likely to be buses running along the road which allow people to get to work.

e This a suitable explanation for the residential sector, using some principles of urban morphology — that is, that major transport routes tend to draw urban development along them. The candidate uses evidence from the map in a sensible way, so the full 3 marks are awarded.

(e) Currently, in Rio de Janeiro, northeast Brazil, urbanisation is occurring which means the population of the city is increasing. Rio is surrounded on three sides by the sea which limits the amount of growth that can occur. The flat land between the mountains and the sea is already built up and full, so new migrants have to build favela settlements on the mountain sides. This has caused deforestation of the hillsides, leading to landslides and mudflows in times of heavy rain. The increase in numbers of people has caused more pollution in the atmosphere and in rivers and streams. The favelas do not have proper sewage systems so the dirty water is left to flow into the rivers. As there are so many people living in Rio, the roads are very congested and there is not much space to build new ones. A new motorway has been built through the mountains to allow people to live further away from Rio in more space and cleaner surroundings.

e Extended answers must refer directly and in detail to actual places. The use of Rio is appropriate, although it would be better if the candidate had located the city correctly in *south*east Brazil! It is clear that the candidate has studied Rio in some detail, as there is mention of the particular physical geography of the site. This is related to environmental problems, such as landslides, in the context of population growth. The answer is encouraging in its content but needs a little more factual detail about Rio itself, such as a couple of names of areas within the city, to be really convincing and so worth the top Level. It can, however, be awarded a Level 2 mark of 7 out of 10.

e **Overall, candidate A scores 19 out of 32 marks, which is equivalent to a low C grade. With greater attention to exam technique, this answer could have been raised a grade.**

AS Geography

Answer to question 5: candidate B

(a) Most of the squatter settlements are located just outside the residential areas, such as just to the west of Magomesi. They can also be clustered close to the industrial areas, such as next to the oil refinery and just south of the CBD. There is a cluster too along Msimbazi Creek.

e This answer is a very good description. A general statement is made with an example to support it. There is also some detail, with actual locations named from the map. The full 4 marks are awarded here.

(b) Squatter settlements are often located near to industrial areas because when people move to the urban area it is usually in search of work and they need to be close to possible jobs.

Squatter settlements are also often located near to main roads and railways. Sometimes, this is because the migrants can set up small shops and businesses to attract passing trade. Also, other people do not want to live next to the noise and pollution of a road or railway, so there is space for the migrants.

They are also located near to sources of water, such as the Msimbazi Creek. Squatter settlements do not have plumbing or sewage works, so the river or creek is used for washing and waste disposal, although this spreads diseases.

e The three reasons given here are different from each other and are all correct. Each reason is clearly stated and then its influence on the siting of squatter settlements is explained. The candidate has used the map to good effect to help with the answers. 3 marks for each reason are awarded, making a total of 9 marks.

(c) Squatter settlements grow in most large urban areas in LEDCs because birth rates are much higher than death rates and so the population increases.

In urban areas in LEDCs, there is frequently not enough planning to cope with the number of people. Often, there are not enough houses, so people have to build their own wherever they can.

e While the first answer is a correct statement, it does not explain why squatter settlements are growing. An increase in numbers need not necessarily bring about large areas of shanty dwellings. The causes are the poverty of the people, who cannot afford anything else, and the inability of the authorities and the housing system to provide enough low-cost houses. No marks are given for the first reason. The second answer is correct, for 3 marks.

(d) The residential sector has developed in a southwesterly direction from the central area as it follows a major road and railway running in this direction. As there is good access along this area, people will want to settle there because they can get to their jobs more easily.

e This is an appropriate reason and well developed. The candidate makes sensible use of the map to support the point and so is awarded the full 3 marks.

OCR (A) Unit 2

(e) Recent changes to Mexico City have caused environmental problems to the area. The high number of people who migrate to this city and the high birth rates mean that the population is growing very rapidly. They don't know exactly how many people are living in Mexico City but it might be as many as 15 million. However, the city is unable to support that many people. The large numbers of squatter settlements are damaging the environment because the sewage is not disposed of and water supplies are polluted. There is also the problem of litter disposal. In areas such as the Vecindades tenements in the centre, there has been a lot of damage from earthquakes which is not always repaired, so the houses are unstable. There is the further problem of air pollution, for which this is one of the worst cities in the world.

> Candidate B's extended answer has some useful points about the actual question asked, such as the issues of water supply and sewage, which are related to the rapid increases in population. It also mentions an actual type of housing particular to the chosen city, while the mention of earthquakes indicates that the candidate has spent some time studying this particular place. As the question asks about *recent* changes, perhaps the point about earthquake damage could have been linked to an actual event. The concern about air pollution needs a little more detail to be really effective. However, this is an encouraging response and it is awarded a Level 3 mark of 8 out of 10.

> **Candidate B scores 27 out of 32 marks, which is a sound A-grade answer.**

AS Geography

Question 6

Urban settlement (II)

Figure 1 shows the factors that influence the spatial residential structures of **MEDC** cities. Family status refers to the ages and stages in the life cycle of individuals and families, and their housing requirements.

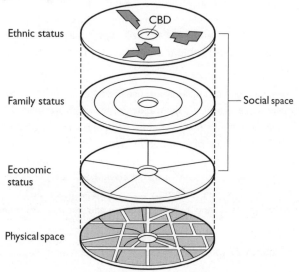

Figure 1

(a) Describe *one* way in which the family status of a neighbourhood might be influenced by the neighbourhood's distance from the city centre. (4 marks)

Table 1 shows the age structure, housing tenure and ethnic composition of three enumeration districts in Preston, Lancashire. The 1:50 000 OS map extract in Figure 2 shows the locations of the enumeration districts.

	District A	District B	District C
Age structure (%)			
0–15 years	37.0	35.3	20.8
16–59 years	60.5	57.0	74.9
60+ years	2.5	7.7	4.3
Housing tenure (%)			
Owner-occupied	79.6	20.3	98.2
Private rented	4.8	0.0	1.8
Public rented	14.3	78.1	0.0
Others	1.3	1.6	0.0
Ethnicity (%)			
White	22.8	99.5	95.1
Non-White	77.2	0.5	4.9

Table 1

OCR (A) Unit 2

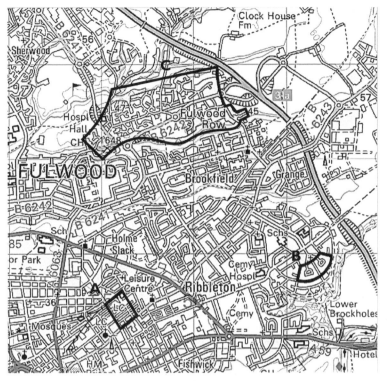

Figure 2

(b) Using the evidence from Table 1 and Figure 2, state and explain *two* possible reasons for the relatively large proportion of children in Districts **A** and **B**. (3+3 marks)

(c) Using the evidence from Table 1 and Figure 2, state and explain *two* reasons for the location of high-income households in District **C**. (3+3 marks)

(d) Suggest a possible explanation for the edge-of-town location of low-income households in District **B**. (3 marks)

(e) Why might an edge-of-town location be more of a disadvantage to residents in District **B** than to those in District **C**? (4 marks)

(f) (i) Using only the evidence of Figure 2, state and explain a possible reason for the large percentage of non-Whites in District **A**. (3 marks)

(ii) Suggest *one* other reason for the large percentage of non-Whites in District **A**. (3 marks)

Total: 29 marks

■ ■ ■

Answer to question 6: candidate A

(a) The inner ring around the CBD is required by those wanting easy access to work in the city centre. Therefore, if an area is near the CBD, its family status will be of working age.

AS Geography

question 6

> 🖉 The candidate does not understand the term 'family status', despite the guidance given in the introductory paragraph. There is a reference to age in the answer, relating this to people working in the CBD, so 1 mark is awarded.

(b) There is a low elderly population in District A (2.5%) and so the population is young and more likely to have children.

There are schools next to District B, so it is easy for children to go to schools near their homes.

> 🖉 The first reason is valid and uses evidence directly from Table 1, so is worth 3 marks. Although the candidate uses Figure 2 for the second reason, the reason is not valid because schools tend to be located throughout the urban area, especially primary schools. No marks are given for the second reason.

(c) It is on the edge of the town and so has plenty of space, which costs more money. Only the high-income people can afford to live here.

There are bigger houses, which cost more, so people with more money can buy houses here.

> 🖉 The candidate correctly mentions the increased space in the first reason, but then repeats this point in the second reason with the comment about 'bigger houses' (which take more space). This is then related to the higher cost of housing. The examiner would read these two answers and take the creditable statements together to give 3 marks out of 6. In these questions, it is best if the answers are quite different from each other.

(d) It is likely to be an overspill local authority council estate. It was built to rehouse people from slum clearance in the inner city.

> 🖉 The candidate has used the data in Table 1 well to identify the housing tenure. The answer, however, only implies the link with low income, so 2 marks out of 3 are awarded. Answers must make a clear link between points in order to be awarded full marks.

(e) Since they rent their houses, the people living in B are unlikely to be able to afford to commute into town easily, because they probably can't afford cars. People in District C are likely to be better off and might own several cars.

> 🖉 This is a good answer as it deals directly with the question and the data provided. The comments about mobility are valid and are worth 4 marks.

(f) (i) Very often, non-Whites are migrants and have less money. They live in the cheaper housing in the inner city, which tends to be terraced and at high density.

> 🖉 The candidate has used the map to identify the type of housing correctly, and has associated this with the economic status of the group. 3 marks are awarded for this answer.

(ii) Most non-Whites travelled by public transport to the centre of the town and so they tended to stay where they first arrived in an area.

e This is not a valid reason to explain the location of this group in District A. The candidate might be muddled about the point that many migrants do end up living in the port of entry to their new country. Factors such as affordable housing and proximity to employment are more convincing ones. No marks can be awarded here.

e **Candidate A scores 16 marks out of a possible 29. This is a C-grade answer.**

■ ■ ■

Answer to question 6: candidate B

(a) If a family has no children, then access to the city centre, where they are likely to work, is important. They don't need much space and can live close to the centre. During child-rearing, space becomes important and so a house in the suburbs, where there is a garden, is preferred.

e This is a well-thought-out response. It shows a good understanding of family status and relates this well to the issue of distance. All 4 marks are awarded.

(b) District A has a high proportion of migrants who may have large families, as their culture is different. They need to live in the cheaper housing found in this district.

In District B, there may be a council estate and this is likely to have many new and young families who cannot afford to buy a house. They are renting a house until they have saved enough money.

e Both these answers contain valid statements based on evidence from the map and the census data. Both explanations make a clear link with children and the relative locations of the two districts. The answers would be strengthened by the addition of some figures from Table 1, but this is not necessary for the response to gain 6 marks. The use of figures can be helpful to an examiner in confirming the candidate's understanding where the written explanation is a little muddled.

(c) The table suggests a large percentage of owner-occupiers. This wealthier group will be able to live in the less urbanised semi-rural environment in the suburbs.

Lots of the houses here are large and can only be bought by high-income people. It is also on the edge of Preston where you need a car to travel around, and high-income people are more likely to have cars.

e Both these reasons have some merit but neither is expressed with sufficient clarity to be worth full marks. Reason 1 correctly identifies the social group living in District C and starts to make a valid association between wealth and an ability to pay for a suburban environment, but it does not go far enough and so receives 2 marks. The second reason starts off with a sentence that would fit better in the first answer. The second sentence introduces a valid factor — the need for increased mobility — and relates this to an ability to pay for this mobility. The phrase 'to travel around' is vague and could be improved by reference to journeys to work, shops or entertainment, so again 2 marks are given. It is poor exam technique to use the word 'also' when the requirement is for one reason only, and it suggests a muddled approach.

(d) District B is mostly local authority housing and this would need to be built on cheap land which is found at the edge. Councils don't have much money and so would not be able to pay very much for the land. People who live in council housing cannot afford to buy their own house and so have to live where the council has houses.

e The candidate makes the correct point about the advantage to local authorities of cheaper land in the suburbs. This is linked with the lower incomes of people who are unable to afford another choice of housing. Full marks are awarded here.

(e) The residents of District B live in council houses. Therefore, they are less likely to have a car since they will be less wealthy. The residents in District C are nearly all buying their own home and so will have a high income. They can afford cars and don't mind travelling from the suburbs.

e Candidate B has made a good point about possible contrasts in car ownership and related this in a straightforward way to the edge-of-town locations. The data in Table 1 are again put to effective use in supporting the answer, with the comment about owner-occupiers. 4 marks are awarded.

(f) (i) When migrants move into an area, they prefer to live where similar migrants already live. They might have family or friends who can help them when they arrive.

e While this is a valid reason and well explained, it cannot be given credit as the question clearly states 'Using only the evidence of Figure 2'. Reading the question with great care, even under timed conditions, is vital. No marks can be given for this answer.

(ii) Migrants are likely to live in cheaper areas as they do not have much money when they first arrive. They look to live in terraced housing, which is usually cheap because it is small and old.

e This is an appropriate reason which is well explained, for 3 marks. It is a point that could have been made in part (i), as there is evidence from the map of terraced housing at relatively high densities in inner Preston, suggesting it is old.

e **In total, candidate B scores 24 out of 29 marks, which is a sound A-grade answer. While some of the responses are not especially convincing in their style, examiners will reward good geography and its application to the resources used in the question.**